Collins *gem*

Kings & Queens

Neil Grant

Consultant: Alison Plowden

HarperCollins Publishers
77-85 Fulham Palace Road
Hammersmith
London W6 8JB

www.collins.co.uk

First published 1996
This edition published 2004

Reprint 11

All photographs supplied courtesy of The Bridgeman Art Library, 17-19 Garway Rd, London, except for those listed on pp. 251-252.

ISBN 13 978 0 00 718885 7

Created and produced by Flame Tree Publishing, part of The Foundry Creative Media Co. Ltd
Crabtree Hall, Crabtree Lane, London SW6 6TW

Special thanks to John Dunne

Printed in China

Contents

Introduction 6
How To Use This Book 8

PART ONE Pre-Saxon Rule 10

Ancient British Chiefs 10
Boudicca 14
The Roman Conquest 16
Roman Rulers 18
Dark Ages 20

PART TWO Scotland, Ireland and Wales 22

Kings of Scots 22
Robert I and the Wars of Independence 40
Flodden 49
The Union of the Crowns 53
Irish Kings 54
The High Kingship of Ireland 58
Princes of Wales 60
Major Welsh Rulers 61
Llywelyn ap Iorwerth, 'the Great' 72

PART THREE Saxons, Normans and Plantagenets 76

Anglo-Saxon Kingdoms 76
The Vikings and the House of Wessex 96
Saxon Kings of England 97

Danish Kings 105
The Normans 114
The Norman Conquest 116
The Angevins 128
The King and the Archbishop 132
The Earl Marshal 138
The Wars of the Roses' 156

PART FOUR Tudors and Stuarts 164

The Tudors 164
The Wives of Henry VIII 170
The Stuarts 178
The Gunpowder Plot 181
The English Civil Wars 184

PART FIVE Hanover to Windsor 196

The Hanoverians 196
The Jacobite Risings 202
The Role of the Monarchy under Victoria 212
Home Rule 214
Parliament and Sovereignty 216
The British Empire 218
The Abdication Crisis 226

PART SIX Compendium 232

The Crown Jewels 232
Nicknames 235
Genealogical Charts 236

Illustration Notes 250
Index 251

Introduction

In many early societies, long before they became organized into states governed by written laws, there existed the figure of the sacrificial king, a figure partly religious, sometimes regarded as a god or as a kind of high priest, partly a symbol representing the spirit of the people, and partly a kind of good-luck token. In some societies, such individuals were actually killed at the end of their appointed reign in a ritual connected with the harvest.

These earliest kings had no power. They were neither rulers nor war leaders but symbols and representatives. There is little connection between the sacrificial king of prehistoric times and the war-making, law-giving kings of medieval Europe, still less with the modern monarchy. Yet this aspect of monarchy, as the symbol of a nation and representative of the people, has survived during the many great changes that have taken place through the centuries. We no longer contemplate cutting a monarch's throat to encourage the crops, but it is essentially that ancient role of national symbol that the monarchy fills today.

As a result of events in the past thousand years or so we have come to regard the monarchy as a political institution: the Queen is, after all, the Head of State. But the once-great political power of the monarchy has been whittled away until practically nothing remains, while the older notion of the monarch as the symbol of national unity is probably as strong as ever it was. Because it is difficult to define, this aspect of monarchy is sometimes underrated or overlooked. It is much easier to describe the Queen's role in the constitution, or her relations with her prime ministers, or her household expenses, yet these are minor matters compared with the almost mystical sense of the monarch as the embodiment of a national idea.

MODERN MONARCHY

That role has little to do with politics or economics. Monarchy cannot be costed and is hard to justify in a modern, democratic society. It is undoubtedly an anachronism, but for many it does seem to fulfil a need.The British monarchy, despite its ups and downs, remains remarkably popular.

SURVIVAL

The British monarchy is not as ancient as the imperial dynasty of Japan, nor quite as old as the Papacy, but it is still one of the most venerable institutions in the world. It has survived many crises, including a brief disappearance in the seventeenth century, and its character has undergone radical change. That it has survived the past two centuries, when crowns were tumbling all around, is partly due to the qualities of recent monarchs. That may be a decisive factor in its future survival.

How To Use This Book

This book details a wide variety of information about Kings and Queens of Scotland, Ireland, England and Wales – everything from facts about the Pre-Saxons through to the Windsors, including concise genealogical charts, to special double pages about issues that characterized some of the most interesting epochs of history. A series of photographs illustrate each topic.

The book is divided into five sections, together with a Compendium at the end.

Each part is colour-coded for easy reference. Part One, which appears in green, presents information about Pre-Saxon rule, including details about ancient British chiefs, Roman rulers and the Roman Conquest. Part Two is colour-coded pink, and provides information about Scotland, Ireland and Wales, with sections on Robert I and the Wars of Independence, The Union of the Crowns, the Princes of Wales and the High Kingship of Ireland. Part Three discusses the Saxons, Normans and Plantagenets and is coded blue. Part Four is coded yellow, with details about the Tudors and Stuarts. Part Five, coded lilac, presents an in-depth discussion of the houses from Hanover to Windsor.

Parts Two to Five provide all the essential information you will need to know about Kings & Queens, including fascinating facts about events that characterized their reigns. You will find details of historical interest and significance, as well as a useful box outlining the key characteristics of each monarch.

The Compendium, coded in lime green, contains useful information about royal regalia, royal nicknames and comprehensive genealogical charts. At the end of the book you'll find an index which lists every monarch found in this book.

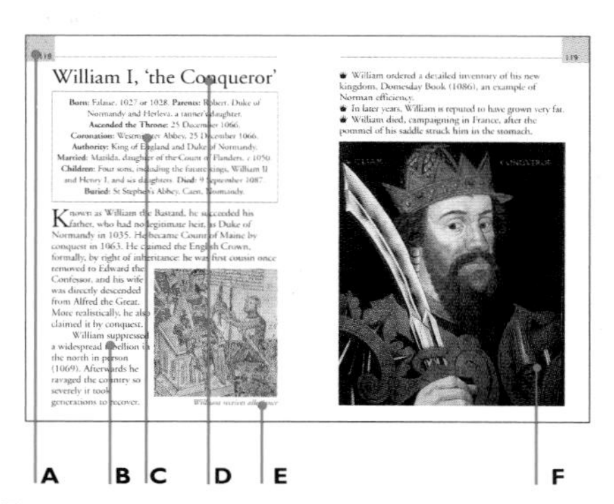
118

William I, 'the Conqueror'

Born: Falaise, 1027 or 1028. **Parents:** Robert, Duke of Normandy and Herleva, a tanner's daughter.
Ascended the Throne: 25 December 1066.
Coronation: Westminster Abbey, 25 December 1066.
Authority: King of England and Duke of Normandy.
Married: Matilda, daughter of the Count of Flanders, *c* 1050
Children: Four sons, including the future kings, William II and Henry I, and six daughters. **Died:** 9 September 1087.
Buried: St Stephen's Abbey, Caen, Normandy.

Known as William the Bastard, he succeeded his father, who had no legitimate heir, as Duke of Normandy in 1035. He became Count of Maine by conquest in 1063. He claimed the English Crown, formally, by right of inheritance: he was first cousin once removed to Edward the Confessor, and his wife was directly descended from Alfred the Great. More realistically, he also claimed it by conquest.

William suppressed a widespread rebellion in the north in person (1069). Afterwards he ravaged the country so severely it took generations to recover.

119

- William ordered a detailed inventory of his new kingdom, Domesday Book (1086), an example of Norman efficiency.
- In later years, William is reputed to have grown very fat.
- William died, campaigning in France, after the pommel of his saddle struck him in the stomach.

A The page number appears in a colour-coded box which indicates which part you are in.

B Essential information about the monarch appears in a concise and fascinating introductory passage.

C A key facts box lists all the important details about a monarch.

D The title of the chapter, in this case the monarch's name, appears at the beginning of every new section.

E A short caption clarifies the subject of an illustration.

F The topic covered on this spread will be illustrated with clear photographs.

PRE-SAXON RULE

Ancient British Chiefs

Knowledge of ancient Britain before the invasion of Julius Caesar in 55 BC depends on archaeological evidence alone. From this we know Britain was closely linked with northern Gaul, by commerce and by kinship, and the main political authorities were the leaders of the Celtic tribes of southern Britain. Traditionally, their centres of power were hill-forts, such as Maiden Castle in Dorset, or groups of scattered habitations, such as Camulodunum (Colchester), which were kingly capitals though not yet towns in the Roman sense.

Some Celtic Tribes of Britain

ATREBATES

This tribe occupied south-central England as well as a section of Belgic Gaul. The leader of the Atrebates in Britain from about 50 BC was **Commius**, who fled to Britain after fighting the Romans in Gaul. He also ruled over the neighbouring Regni (who occupied what is today West Sussex). His lands were split up under his sons, **Tincommius, Eppillus** and **Verica.**

Tincommius adopted a pro-Roman policy and was recognized by the Emperor Augustus as a 'client-king' about 15 BC. Like his brother Verica, who issued coins bearing a vine-leaf motif and ruled the area of the Regni, he came under pressure from Cunobelinus (see below) and eventually fled to Rome.

Ancient Britons acknowledging Caesar .

CATUVELLAUNI

The homeland of the Catuvellauni was modern Hertfordshire. Their king, **Cassivellaunus** (Caswallon), took the lead in resisting the invasion of Julius Caesar. He overcame the customary tribal animosities of Celtic Britain by forming an alliance of neighbouring tribes. His headquarters was Verulamium (St Albans). His successor, **Tasciovanus,** was recognized as a Roman client-king at the same time as Tincommius. Under his son, **Cunobelinus** (Shakespeare's Cymbeline), the Catuvellauni became the leading power in southern Britain before the Roman Conquest.

Cunobelinus conquered the Trinovantes (see below) early in the first century AD, gaining Camulodunum, the chief centre for continental trade, which became his capital. He took over part of Kent, the Midlands and territory of the Atrebates in Sussex and Hampshire. Coins minted in Camulodunum described him as rex 'king' and he was referred to as 'king of the Britons'.

Outlying provinces were governed by sons of Cunobelinus or other members of his family. His personal empire ended with his death. His son **Togodomnus** was killed fighting the Romans. Another son, **Caratacus** (Caradoc), who ruled from the old Atrebates settlement of Calleva (Silchester), retreated to Wales. Caratacus led the Welsh tribes, the Silures and Ordovices, against the Romans until AD 50 when, defeated, he retreated to the territory of the Brigantes (see below) who betrayed him to Rome. A third son of Cunobelinus, **Adminius,** ruled in Kent until he fled to Rome in about AD 40.

TRINOVANTES

The Trinovantes, whose kings in the pre-Roman period included **Addedomarus, Tasciovanus** and **Dumnovellaunus,** formed an alliance with Julius Caesar during his invasions of 55 and 54 BC. Caesar protected them against their powerful and aggressive neighbours, the Catuvellauni. Fifty years later Roman policy had changed: political entanglements outside the empire were avoided, and Cunobelinus was able to subjugate not only the Trinovantes but also other friends of Rome, including the Atrebates.

ICENI

The Iceni were an East Anglian tribe, northern neighbours of the Trinovantes. At the time of the Roman Conquest they were ruled by a king named **Prasutagus,** who was one of the few British rulers allowed some autonomy as a client-king after the Conquest. On his death, dispute over the succession provoked a major revolt against Roman rule led by his widow **Boudicca** (Boadicea).

BRIGANTES

The Brigantes was probably the largest tribe in Celtic Britain, though comprising a loose federation rather than a centralized community. They occupied most of modern-day England north of the Mersey and the Humber. Their chief settlement was initially at Isurium (Aldborough) and later at Eboracum (York). Their queen, **Cartimandua,** who reigned about AD 50-70, was responsible for turning the fugitive Caratacus over to the Romans, but the Brigantes were to prove unreliable subjects of Rome for the next 100 years.

Julius Caesar

Boudicca

When Prasutagus, friendly client-king of the Iceni, died, the Romans confiscated his property. When his widow Boudicca (Boadicea) protested, the Romans had her flogged and her daughters raped. The Trinovantes also had been incensed by Roman brutality. Overbearing Roman colonists had seized their lands. Boudicca led a rebellion (AD 61). She was joined by the Trinovantes, also former friends of Rome, and others. The rebels massacred Roman settlements, destroyed a small force sent to suppress them, and sacked Camulodunum, Verulamium and Londinium. Romans were tortured and massacred. The governor, Suetonius Paulinus, met the rebels somewhere in the Midlands and defeated them. About 80,000 Britons are said to have died.

- When Boudicca's revolt broke out, the main Roman army was in North Wales. It had just captured Anglesey, stronghold of the Druids, who had been encouraging resistance to Rome, and had to march halfway across England to defeat Boudicca.
- Boudicca died after her defeat by Paulinus, probably by taking poison.

Iceni coins found at Eriswell, Suffolk

The Roman Conquest

JULIUS CAESAR

Julius Caesar (*c.*102-44 BC) made two raids on southern Britain, in 55 and 54 BC respectively. Caesar himself testified that his object was to prevent the Celts of southern Britain giving aid to their relations in northern Gaul, which he was in the process of conquering. The effect of his raids was to draw southern Britain into the Roman world.

THE INVASION OF CLAUDIUS

Owing to other distractions, the Roman Conquest was delayed nearly a century. In AD 43, 50,000 Romans under Aulus Plautius landed at Richborough and other places. Emperor Claudius (10 BC to AD 54) made a 16-day visit two months later, by which time the Romans had already reached the Thames. He witnessed the capture of Camulodunum, where the attacking Roman force included elephants. Some tribes resisted strongly. Others, hostile to the hegemony of the Catuvellauni, quickly surrendered. Most of lowland Britain was swiftly overrun.

ADVANCE OF THE LEGIONS

The legions fanned out from Camulodunum. Emperor Vespasian (AD 9-79) defeated the Durotriges in Dorset, capturing Maiden Castle after a fierce battle. Other forces penetrated the Midlands.

- The Romans advanced on strongpoints like Maiden Castle in a phalanx called a 'tortoise'. Men in the centre held their shields above their heads, giving all-round protection against missiles.
- In the wake of the Conquest, forts and barracks were built at strategic points. Road-building began, to speed the movements of troops.
- Initially, the Romans allowed compliant local rulers to remain in charge, under the governor, who was also commander of the troops.

Head of Caesar Augustus, the first Roman Emperor

Roman Rulers

The Romans conquered nearly all of Britain, including Caledonia (Scotland). Roman settlement was restricted to lowland Britain. The remainder, including Wales and southern Scotland, was under military control.

Britain became a Roman province (subdivided into two in the third century, with capitals at Londinium [London] and Eboracum [York], and into four in the fourth century, capitals at London, York, Cirencester and Lincoln). It was ruled by a governor under the Emperor. The country was divided into districts, locally administered, corresponding roughly to the old tribal territories. Some estates were held directly by the Emperor. The Celtic upper class in lowland Britain became thoroughly 'Romanized'.

There were frequent attacks from unconquered tribes, especially in the north. **Emperor Hadrian** (117-138), who visited Britain in 122, ordered the building of Hadrian's Wall between the mouth of the Tyne and the Solway Firth, marking the northern frontier of the empire. **Antoninus Pius** (emperor 138-61) extended the boundary northward, building the Antonine Wall between the Firths of Forth and Clyde.

Emperor **Septimus Severus** (193-211), with his son Caracalla, spent the last three years of his life in Britain, suppressing unrest and leading a punitive expedition to Caledonia. He died at Eboracum (York).

From the late second century, the Roman Empire

was often on the verge of disintegration. Army discipline broke down. During the numerous contests by generals aspiring to become emperors, several were proclaimed in Britain. In 259-73 Britain was part of a breakaway empire that included Gaul. At least 25 emperors ruled between 235 and 284. All but one suffered violent deaths.

In 287 a rebellion was led by a Roman naval commander in Britain, **Marcus Aurelius Carausius,** who controlled both sides of the English Channel. He was murdered in 293 by his successor, **Caius Allectus.**

Constantius Chlorus I (293-306) defeated Caius Allectus in 296. He also died at Eboracum. His son, **Constantine the Great**, re-established imperial authority. Constantine founded a new capital, Byzantium, which resulted in the division of the empire into west and east. The western empire came under continual attack from Germanic tribes.

Roman Britain was prosperous during the 4th century, but under increasing attack from neighbours, from the Low Countries including Germanic tribes such as the Saxons. Defensive forts were built along the south-eastern coast: the 'Saxon Shore'.

The Roman occupation lasted from AD 43 to about 400. Most troops had been withdrawn earlier to deal with trouble in other parts of the empire. In 410 Emperor **Honorius** (384-423) told the Britons to expect no further help from Rome.

The Dark Ages

*c.*425-50 Vortigern British high king
446 British appeal to Aetius, Roman commander in Gaul, for help against Saxons
*c.*449 Revolt of Hengist and Horsa
460 Celts defeat Jutes at Richborough
473 Hengist defeats Celts
*c.*495 West Saxon kingdom founded
*c.*517 Battle of Badon
*c.*538 Arthur killed at battle of Camlann

In the early fifth century Britain was under attack not only from Saxons and other Germanic tribes from the east, but also from Picts, Scots and Irish tribes.

The chief ruler, or high king, in southern Britain was called **Vortigern** (actually his title, not name). Vortigern hired Germanic mercenaries, led by the brothers **Hengist** and **Horsa,** in the 440s, offering land in Kent in return for service against northern raiders. Ambitious for more land, the brothers turned against the Britons. Though defeated by Vortigern's son, **Vortimer** (who was killed during the battle) in 460, they soon returned; Hengist and his son **Aesc** established the kingdom of Kent, the first Anglo-Saxon kingdom, after defeating the Britons in 473.

THE JUTES

Hengist and Horsa and their men were called 'Jutes' by the chronicler Bede. Their home was, however, probably not Jutland, as Bede thought, but the Rhineland, where they were closely related to the Angles and Saxons.

After the unsuccessful Vortigern, a new leader emerged, **Ambrosius Aurelianus,** identified with a Welsh hero, **Emrys.** He may have replaced Vortigern as a kind of high king, uniting the Britons against the invaders. He is said to have been a great cavalry commander, who held the Jutes in check and won a victory over the Angles in 493.

Emrys's successor (perhaps his son) is more famous, but as a mythical hero, not a historical figure. This was **Arthur**, said to have fought twelve battles against invaders throughout Britain, including his last and greatest in which the Saxons were routed. Arthur's death, probably in a conflict among the Britons, left a divided people without a leader.

Scotland, Ireland & Wales

Kings of Scots

THE ROMANS

Gnaeus Julius Agricola (AD 40-93), Roman governor of Britain, invaded Caledonia (Scotland) in AD 81. He defeated the Celtic tribes under their leader, Calgacus, at the battle of Mons Graupius (AD 83), in eastern Scotland. The difficult country and fierce resistance of the tribes convinced the Romans that Caledonia could not be conquered and held. Emperor Hadrian ordered the building of a defensive wall (122) between the Solway and the Tyne. A few later sorties notwithstanding, Hadrian's Wall marked the northern limit of the Roman Empire.

In the fifth century Scotland was divided between four peoples:

The **Picts,** who were Celts, were the strongest, controlling the country from the Forth to Caithness.

Another Celtic people inhabited **Strathclyde,** from Cumbria to the Solway.

The **Scots** occupied the

Saxon warrior

kingdom of Dalriada in Kintyre, Argyll and neighbouring isles. They were Celts who came from northern Ireland in about the third and/or fourth centuries. Their kings were to establish what became the Scottish monarchy.

South-east Scotland was occupied by Anglo-Saxons, who created the kingdom of **Northumbria** in the early seventh century.

HOUSE OF FERGUS, OR ALPIN

Fergus was the semi-legendary chief of the Scots of Dalriada, said to have established the capital at Dunadd, near Crinan, *c.*500. His successors before the ninth century are obscure.

The succession of early Scottish kings is complicated by the custom of passing the Crown between different branches of the dynasty in succeeding generations. In an age when kings frequently died by violence comparatively young, this custom ensured that the Crown was likely to be inherited by a mature man rather than a child.

Kenneth Mac Alpin (died 858), regarded as the founder of the monarchy, was the son of Alpin, King of Dalriada, and a Pictish princess. He succeeded his father in 834 and, after defeating the Picts in 843, united virtually all of Scotland north of the Forth in his kingdom of Alba, or Alban.

Kenneth Mac Alpin founded the ecclesiastical capital at Dunkeld and made his secular capital at Forteviot. He had at least five children. His two sons were later kings. His three daughters married, respectively, the King of Strathclyde; the Norse King of Dublin, Olaf the White; and the High King of Ireland, Aedh Finnlaith.

Donald I (reigned 858-62), brother of Kenneth Mac Alpin, possibly died in battle (few of these Kings of Scots died peacefully) against the Norsemen.

Constantine I (reigned 862-77), son of Kenneth Mac Alpin, died in battle against the Norsemen.

Aedh (reigned 877-78), another son of Kenneth Mac Alpin, died in battle, possibly buried at Maiden Stone, Aberdeen.

Eochaid (reigned 878-89), son of Run Macarthagail, King of Strathclyde, whose wife was a daughter of Kenneth Mac Alpin. Through his father, he was also King of Strathclyde. He was deposed shortly before his death and apparently had no issue.

Donald II (reigned 889-900), only son of Constantine I. He was killed in battle and buried, like most early kings of Scots, on the holy isle of Iona.

Constantine II (reigned 900-42, died 952), son of Aedh. He acknowledged the English king Edward the Elder as overlord, and was defeated by the English at the battle of Brunanburgh (937). He abdicated in order to become a monk at St Andrews, subsequently being elected abbot.

Malcolm I (reigned 942-54), son of Donald II, was killed in a battle with the men of Moray and buried at Iona.

Indulf (reigned 954-62), son of Constantine II. Like his father, he abdicated in order to enter a monastery, but shortly afterwards he was killed during a Norse raid.

Duff (reigned 962-66/7), son of Malcolm I, died in battle.

Colin (Cuilean) (reigned 966/7-71), son of Indulf, was killed in battle against the King of Strathclyde.

Kenneth II (reigned 971-95), son of Malcolm I, was murdered by supporters of his successor.

Constantine III (reigned 995-97), son of Colin, was killed, probably murdered, at Rathinveramon.

Kenneth III (reigned 997-1005), son of Duff, was killed in battle against his successor.

Malcolm II

Born: *c.*954. **Parents:** Kenneth II and a Leinster princess. **Ascended the Throne:** 25 March 1005. **Reign:** 18 years. **Married:** Name unknown. **Children:** Two or three daughters, including Bethoc, the Lady of Atholl. **Died:** Glamis, 25 November 1034.

The last of the House of Alpin, Malcolm II defeated the Anglo-Saxons at Carham on the River Tweed (1018), thus gaining Lothian. He also had a claim to the kingdom of Strathclyde, which he secured for his grandson and heir, Duncan, when the reigning King of Strathclyde died without issue.

THE HOUSE OF DUNKELD

Since Malcolm II had no male heir, the succession passed to the House of Dunkeld through the marriage of his daughter, Bethoc, to Crinan, lay Abbot of Dunkeld. The dynasty lasted until 1290. Its members were somewhat less prone than their predecessors to violent and early death.

Duncan I

Born: *c.*1001. **Parents:** Crinan (Cronan), Mormaer of Atholl and lay Abbot of Dunkeld, and Bethoc, daughter of Malcolm II. **Ascended the Throne:** November 1034. **Reign:** Six years. **Married:** Sybilla, daughter or sister of the Earl of Northumbria. **Children:** Three sons, including Malcolm III and Donald III, and probably one daughter. **Died:** 1 August 1040.

Duncan's reign was short and unsuccessful. He was killed, probably in battle at Pitgaveny near Elgin, by a rival claimant and cousin, Macbeth.

Macbeth (Maelbeatha)

Born: *c.*1005. **Parents:** Finlay MacRory, Mormaer of Moray, and Donada, daughter of either Kenneth II or Malcolm II. **Ascended the Throne:** 14 August 1040. **Reign:** 17 years. **Married:** Gruoch (Gruach), a granddaughter of Kenneth III. **Children:** None. **Died:** 15 August 1057.

Macbeth was unfairly treated by Shakespeare, his reign being longer and his rule more capable than the English playwright suggested. He claimed the throne through his wife. Macbeth is said to have made a pilgrimage to Rome. He was defeated at Dunsinane, near Scone in Perthshire, in 1054, and later killed at the Battle of Lumphanan near Aberdeen against the future Malcolm III, son of Duncan, and his English allies led by Earl Siward of Northumbria.

Lulach

Born: *c.*1030. **Parents:** Gillacomgain, Mormaer of Moray, and Gruoch, who married Macbeth as her second husband. **Ascended the Throne:** 15 August 1057. **Crowned:** Scone, August 1057. **Reign:** Seven months. **Married:** A daughter of the Mormaer of Angus. **Children:** One son, one daughter. **Died:** Essier, Strathbogie, 17 March 1058.

Macbeth's stepson and successor was the first King of Scots known to have taken part in a coronation ritual at Scone. Lulach was killed by Malcolm III less than a year after succeeding Macbeth.

Macbeth instructing murderers to kill Banquo

Malcolm III

Born: *c.*1031. **Parents:** Duncan I and Sybilla of Northumbria.
Ascended the Throne: 17 March 1058.
Reign: 35 years. **Married:** (1) Ingibiorg, widow of Earl Thorfinn II of Orkney, (2) Margaret, daughter of Edward the Atheling and granddaughter of King Edmund II of England.
Children: With (1), three sons, including the future Duncan II; with (2), seven sons, including three future kings, and two daughters. **Died:** Alnwick, 13 November 1093.
Buried: Tynemouth.

Malcolm III was brought up in England from the age of nine, and in 1069 married an English princess, later canonized as St Margaret for her patronage of the Church.

The couple introduced a strong English influence into both lay and ecclesiastical society. The centre of gravity of the kingdom moved south, into Anglo-Saxon Lothian and away from the Celtic north.

After the Norman Conquest of England, Malcolm supported the claims of Margaret's brother, Edgar the Aetheling, to the English throne and launched a series of raids into northern England. After a retaliatory invasion by William the Conqueror, Malcolm paid homage to him at Abernethy in 1071. He was ambushed and killed while besieging Alnwick on his final Northumbrian raid. Margaret died a few days later.

King Malcolm III

Donald III

Born: *c.*1033. **Parents:** Duncan I and Sybilla of Northumbria. **Ascended the Throne:** 13 November 1093. **Reign:** Four years. **Married:** Unknown. **Children:** One daughter, Bethoc. **Died:** Rescobie, Forfar, 1099. **Buried:** Dunkeld, later removed to Iona.

Donald Ban, or Donaldbane, seized the throne on the death of his brother, Malcolm III. His background was Celtic and Norse and he reversed Malcolm's Anglo-Norman policies. Attacked by Malcolm's sons, with English support, he lost the throne (1094), regained it, but was again deposed (1097), dying in captivity. He was the last member of the House of Dunkeld to be buried in Iona, which subsequently fell to the Norsemen.

Duncan II

Born: *c.*1060. **Parents:** Malcolm III and Ingibiorg. **Ascended the Throne:** May 1094. **Reign:** Six months. **Married:** Ethelreda, daughter of Earl Gospatrick of Northumbria. **Children:** One son, William. **Died:** 12 November 1094. **Buried:** Dunfermline Abbey.

Duncan II overthrew his uncle, Donald III, with English help but Donald regained the throne when Duncan was killed in battle or, perhaps, murdered.

Edgar

Born: *c.*1073. **Parents:** Malcolm III and Margaret.
Ascended the Throne: October 1097. **Reign:** Nine years.
Married: Unmarried. **Children:** None.
Died: 8 January 1107. **Buried:** Dunfermline Abbey.

Helped to the throne by the English, Edgar admitted many Anglo-Norman settlers to Scotland and acknowledged William II of England as overlord. He formally ceded Kintyre and the Hebrides to King Magnus 'Barelegs' of Norway, who already occupied those territories.

King Edgar

Alexander I

Born: *c.*1078-80. **Parents:** Malcolm III and Margaret.
Ascended the Throne: 8 January 1107.
Reign: 17 years. **Married:** Sybilla, illegitimate daughter of Henry I of England. **Children:** No legitimate children.
Died: Stirling Castle, April 1124. **Buried:** Dunfermline Abbey.

Alexander I ruled over a limited area, between the Forth and the Spey, making no attempt to control Ross and Argyll and leaving the south to his brother and successor, David I. The English connection grew stronger: Henry I was Alexander's father-in-law and brother-in-law (being married to Alexander's sister) as well as his overlord.

King David I

David I

Born: *c.*1084. **Parents:** Malcolm III and Margaret.
Ascended the Throne: April 1124.
Reign: 29 years. **Married:** Matilda, daughter of the Earl of Huntingdon. **Children:** Two sons, two daughters.
Died: Carlisle, 24 May 1153.
Buried: Dunfermline Abbey.

The last and most able of the sons of Malcolm III, David I ruled southern Scotland on behalf of his brother Alexander I before his accession. He had been brought up in England and, as Prince of Cumbria and Earl of Huntingdon (through his wife), he was one of the most powerful of English barons as well as King of Scots. English influence was consolidated during David's long and unusually peaceful reign. A feudal system headed by Anglo-Norman barons was established, trade and urban life encouraged, and royal government strengthened. The king supported the Church, founding bishoprics and parishes and endowing abbeys such as Melrose and Kelso. David's intervention in the English civil wars of Stephen and Matilda led to the acquisition of part of Northumbria.

Royal authority hardly existed in the Highlands, and the Western Isles were still nominally Norwegian.

Malcolm IV

Born: 20 March 1141. **Parents:** Henry, Earl of Northumberland, and Ada de Warenne. **Ascended the Throne:** 12 June 1153. **Reign:** Twelve years. **Married:** Unmarried. **Died:** Jedburgh, 9 December 1165. **Buried:** Dunfermline Abbey.

Grandson of David I, Malcolm IV came to the throne as a boy of eleven, and was soon forced to cede his grandfather's conquests in northern England to the powerful Henry II. Nor could he prevent the King of Norway sacking Aberdeen, or Somerled, ancestor of the Macdonald Lords of the Isles, sacking Glasgow.

David I and Malcolm IV

William I

Born: *c.*1143. **Parents:** Henry, Earl of Northumberland, and Ada de Warenne. **Ascended the Throne:** 9 December 1165. **Reign:** 49 years. **Married:** Ermengarde de Beaumont, daughter of an illegitimate daughter of Henry I of England. **Children:** One son, the future Alexander II, and three daughters. **Died:** Stirling Castle, 4 December 1214. **Buried:** Arbroath Abbey.

A more formidable character than his brother, Malcolm IV, William I, known as the Lion, concluded an alliance with France – the beginning of the 'Auld Alliance' – against England. Defeated and captured, William 'the Lion' was forced to accept humiliating terms from the English in the Treaty of Falaise (1174), but later retrieved Scottish independence by agreement with Richard I of England, initiating nearly a century of peace between the two kingdoms.

He failed to subjugate the rebellious Celtic south-west, or to assert his authority over the MacDougall Lords of Lorne and the Macdonald Lords of the Isles.

Alexander II

Born: 24 August 1198. **Parents:** William I and Ermengarde de Beaumont. **Ascended the Throne:** 4 December 1214. **Crowned:** Scone, 6 December 1214. **Reign:** 35 years. **Married:** (1) Joan, daughter of King John of England, (2) Marie de Coucy. **Children:** One son, the future Alexander III; one illegitimate daughter. **Died:** Isle of Kerrara, 6 July 1249. **Buried:** Melrose Abbey.

On his accession to the throne at the age of sixteen, Alexander was approached by the English barons and asked for his support in their campaign against King John. He led an army across the border into the Northern districts and harried the King's supporters. After the death of John, Alexander continued to support struggles against the young Henry III – in particular the campaign led by Prince Louis of France. In 1237, however, Alexander agreed terms with Henry in the Treaty of York, where the Scottish king abandoned his house's claim to Northumbria in exchange for some English estates. The other notable success of his reign was his expedition against the dissidents in Argyllshire and the islands, where the inhabitants were known to be unsupportive of the Scottish rule and to encourage rebellion. His efforts to assert royal authority in the west, though, had limited success and were ended by his death on campaign.

Alexander III

Born: 4 September 1241. **Parents:** Alexander II and Marie de Coucy. **Ascended the Throne:** 8 July 1249. **Crowned:** Scone, 13 July 1249. **Reign:** 37 years. **Married:** (1) Margaret, daughter of Henry III of England, (2) Yolande, daughter of the Comte de Dreux (1285). **Children:** With (1), Margaret, Alexander and David. **Died:** Near Burntisland, Fife, 19 March 1286. **Buried:** Dunfermline Abbey.

Alexander III continued his father's efforts to establish royal authority in the west. In 1263 he defeated Haakon of Norway at the Battle of Largs (1263), and subsequently ended for ever the centuries-old rivalry between the royal houses of Scotland and Norway, although the Hebrides, while exchanging overlords, remained virtually independent under the Lords of the Isles. Otherwise, his reign was a period of comparative peace and growing prosperity.

Riding towards Kinghorn in Fife at night, his horse fell and he died from his injuries. All his children predeceased him. At his death, his heir was his three-year-old granddaughter Margaret, the 'Maid of Norway.'

Margaret

Born: Early in 1283. **Parents:** King Erik II of Norway and Margaret, daughter of Alexander III. **Ascended the Throne:** 19 March 1286. **Reign:** Four years, five months. **Died:** At sea, September 1290. **Buried:** Bergen, Norway.

The 'Maid of Norway', betrothed to the son of Edward I of England, died on the voyage from Norway to Scotland. She was the first Queen Regnant of Scots, and the last member of her dynasty.

FIRST INTERREGNUM

The death of the Maid of Norway left Scotland without an obvious heir to the throne. In this political vacuum, the most important figure was the powerful King Edward I of England. In 1290 he was asked to decide who should be king from among 13 competitors. Of the two outstanding candidates, both of them Anglo-Norman lords who had fought in the English army, Edward chose John Balliol as likely to prove more amenable than his rival, Robert le Brus.

◀ *Robert Bruce, from the Seton Armorial*

John

Born: *c.*1249. **Parents:** Hugh de Balliol of Barnard Castle and Devorguilla of Galloway, a great-granddaughter of David I. **Ascended the Throne:** 17 November 1292. **Crowned:** Scone, 30 November 1292. **Reign:** Four years. **Married:** Isabella de Warenne, a granddaughter of King John of England. **Children:** Two sons, Edward and Henry, and one or two daughters. **Died:** Normandy, 1313.

John Balliol, reputedly a weak character under the dominance of his English overlord, Edward I, rebelled in 1296 and, fortified by a French alliance, invaded England. Edward launched a counter-invasion, supported by Bruce, later Robert I, and others of his Scottish vassals. Being defeated, Balliol abdicated and, after a spell in prison, retired to his estates in Normandy.

SECOND INTERREGNUM

Edward conquered Scotland and extracted homage from the chief landholders, who acknowledged him as king at Berwick in 1296. National resistance broke out in 1297, led by William Wallace. Captured in 1305, Wallace was executed, but the struggle for independence was renewed when some of Edward's chief Scottish vassals, including Robert Bruce, turned against him.

Robert I and the Wars of Independence

Born: Writtle, Essex, 11 July 1274. **Parents:** Robert le Brus, Lord of Annandale, and Margaret (Marjorie), daughter of Neil, Earl of Carrick. **Crowned:** Scone, 27 March 1306. **Reign:** 23 years. **Married:** (1) Isabella, daughter of the Earl of Mar, (2) Elizabeth de Burgh, daughter of the Earl of Ulster. **Children:** With (1), a daughter, Marjorie; with (2), two sons, including David II, and two daughters. **Died:** Cardross Castle, 7 June 1329. **Buried:** Dunfermline Abbey (his heart at Melrose Abbey).

In 1306 Robert Bruce and Red John Comyn, rival claimants to the Scottish Crown, met in a church in Dumfries. They quarrelled and Bruce killed Comyn. The act antagonized both the powerful Comyn family and the Church. Bruce nevertheless had himself crowned king at Scone. His position, already weak, became desperate when Edward I sent an army which defeated him at Methven. He became an outlaw on the run.

The tomb of Robert Bruce

SCOTTISH RECOVERY

A later legend relates how Bruce, hiding out in a cave, was inspired to renew resistance to the English by watching a spider, which, having tried and failed to attach its thread to a beam six times, refused to give up and was rewarded by success on the seventh attempt.

In spite of further defeats, his supporters, including leaders of future clans – Donald, Campbell and Maclean – increased. After Edward I's death in 1307, his weaker son, Edward II, abandoned the Scottish campaign. Bruce was able to retake English-held strongpoints and even invade England. The decisive act was played at Bannockburn, near Stirling, on 24 June 1314, when Bruce won a dramatic victory over Edward II's invading army.

The English knights were bogged down in marshy ground beside the Bannock burn, with the Scots above them. Edward's army fled, leaving the burn choked with English corpses.

The war dragged on for 14 years, with the Scots now on the offensive. By the Treaty of Northampton (1328), the young Edward III abandoned English claims on Scotland.

Bruce at Bannockburn

David II

Born: Dunfermline, 5 March 1324. **Parents:** Robert I and Elizabeth de Burgh. **Ascended the Throne:** 7 June 1329. **Crowned:** Scone, 24 November 1331. **Reign:** 41 years, with interruptions. **Married:** (1) Joan, daughter of Edward II of England, (2) Margaret Drummond. **Children:** None. **Died:** Edinburgh Castle, 22 February 1371. **Buried:** Holyrood Abbey.

The only surviving son of Robert Bruce, David II was driven into exile aged ten by Edward Balliol in 1334. Robert Stewart, Bruce's 17-year-old grandson, upheld his cause against Balliol and the English. David returned from France in 1341 and, responding to a French appeal for help, invaded England in 1346 and was captured at the Battle of Neville's Cross. He remained a prisoner in the English court until the Treaty of Berwick (1357) restored him in exchange for a ransom and a promise to make an English prince his heir.

Edward

Born: Unknown. **Parents:** King John Balliol and Isabella de Warenne. **Crowned:** Scone, 24 September 1332. **Reign:** Four years, with interruptions. **Married:** Unmarried. **Died:** Probably at Wheatley, Yorkshire, January 1364.

Edward Balliol was the English candidate in the complex Scottish-English-French power struggle. He was deposed by the adherents of David II in December 1332, restored in 1333, deposed again in 1334, restored in 1335 and finally deposed in 1341. He left his claim to the Scottish Crown to Edward III of England.

THE HOUSE OF STEWART

The Scots rejected David II's promise to hand over the kingdom to the English, and the Crown passed to the former regent, Robert Stewart, whose surname derived from his father's office of High Steward of Scotland. Eventually, instead of English kings ruling the Scots, Stewart kings were to reign in England, though the members of this long-lasting dynasty were of highly variable monarchical abilities. Many of the early Stewart kings succeeded as young children, creating a power vacuum exploited by the great nobles.

Edward's royal seal

Robert II

Born: Paisley, 2 March 1316. **Parents:** Walter, High Steward of Scotland, and Marjorie, daughter of Robert I. **Ascended the Throne:** 22 February 1371. **Crowned:** Scone, *c.*26 February 1371. **Reign:** 19 years. **Married:** (1) Elizabeth Mure, (2) Euphemia, daughter of the Earl of Ross. **Children:** With (1), four sons and six daughters; with (2), two sons and two daughters; eight or more illegitimate children. **Died:** Dundonald Castle, 19 April 1390. **Buried:** Scone.

Robert II was less effective as king than he had been as regent for David II. Conflict between Crown and nobility became a major disruptive force in Scottish government, lasting for centuries.

Robert III

Born: *c.*1337 (christened 'John').
Parents: Robert II and Elizabeth Mure. **Ascended the Throne:** 19 April 1390. **Crowned:** Scone, 14 August 1390. **Reign:** 16 years. **Married:** Annabella Drummond of Stobhall. **Children:** Three sons and four daughters; one illegitimate son. **Died:** 4 April 1406. **Buried:** Paisley Abbey.

Robert had been crippled by a horse and virtually abdicated in 1399. The regency was disputed between his son, David, Duke of Rothesay, and his brother Robert, Duke of Albany. Rothesay was kidnapped and, probably, murdered in 1402, leaving Albany supreme.

James I

Born: Dunfermline, December 1394.
Parents: Robert III and Annabella Drummond.
Ascended the Throne: 4 April 1406. **Crowned:** Scone, May 1424. **Reign:** 20 Years. **Married:** Joan (or Jane) Beaufort, daughter of the Earl of Somerset and great-granddaughter of Edward III of England. **Children:** Two sons, including James II, and six daughters. **Died:** Perth, 21 February 1437. **Buried:** Perth.

James was captured by the English in 1406 and held hostage until ransomed in 1424. During his absence, the Duke of Albany and, after his death in 1420, his son Murdoch, held power as regents.

During the regency the great nobles built up their power and independence. The Macdonald Lords of the Isles maintained total autonomy. Invading Islesmen sacked Aberdeen in 1411 but retreated after a bloody battle with Albany's allies at Harlaw.

James's vigorous but vindictive reforming policies made him unpopular, and he was assassinated by kinsmen.

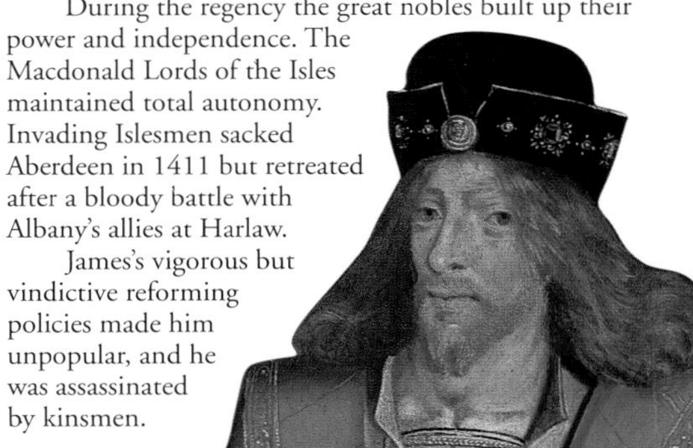

James II

Born: Holyrood Abbey, 16 October 1430.
Parents: James I and Joan Beaufort.
Ascended the Throne: 21 February 1437.
Crowned: Kelso Abbey, 25 March 1437.
Reign: 23 years. **Married:** Mary, daughter of the Duke of Guelders. **Children:** Four sons, including James III, and two daughters. **Died:** Roxburgh, 3 August 1460.
Buried: Holyrood Abbey.

After another troublesome minority, James II assumed control in 1449 and took strong action to quell the disruptive nobility. The mighty Douglasses were broken when their leaders were invited to dinner with the young king in 1440 and murdered. Their successor was killed by James himself in 1452. While besieging Roxburgh, held by the English, a gun exploded and killed him.

James III

Born: St Andrews (?Stirling), probably in May 1452. **Parents:** James II and Mary of Gueldres. **Ascended the Throne:** 3 August 1460. **Crowned:** Kelso Abbey, 10 August 1460. **Reign:** 28 years. **Married:** Margaret of Denmark, daughter of the King of Denmark, Norway and Sweden. **Children:** Three sons, including James IV. **Died:** Milltown, near Stirling, 11 June 1488. **Buried:** Cambuskenneth Abbey.

The dowry of James III's wife, Margaret, included sovereignty of the Northern and Western Isles. After another regency, James took control in 1469. He was more of a scholar than a warlord, and his efforts to assert royal authority provoked rebellions and an English invasion. His brothers, whom he had imprisoned, joined his opponents. One, the Duke of Albany, was proclaimed King of Scots after escaping to London. A later conspiracy set up James's own son as king. James confronted the rebels at Sauchieburn (1488) and was hurt during the battle when his horse threw him. A man masquerading as a priest stabbed him to death.

James IV

Born: 17 March 1473. **Parents:** James III and Margaret of Denmark. **Ascended the Throne:** 11 June 1488. **Crowned:** Scone, 26 June 1488. **Reign:** 25 years. **Married:** Margaret Tudor, daughter of Henry VII of England. **Children:** Four sons, including James V, and two daughters; seven illegitimate children. **Died:** Flodden Field, 9 September 1513. **Buried:** Possibly Sheen Abbey, Surrey; his head, possibly, in St Michael's Church, Wood Street, London.

The most distinguished member of his dynasty, James IV was a Renaissance prince, a patron of the arts, energetic, intelligent – he spoke Gaelic among other languages – and a born leader. For the first time in a century, there was no minority. James crushed the rebels and re-established royal authority, presiding over unprecedented economic growth, educational and artistic development and cultivated town life. The Lordship of the Isles was ended but James's goodwill visit in 1494 had disappointing results, so he resorted to tougher policies, relying on feudal magnates like the Campbell Earl of Argyll. He was killed in the Battle of Flodden Field.

Flodden

When France was attacked by a European coalition that included England, she appealed for help to her only ally, Scotland. James IV realized that the defeat of France would put Scotland in danger. He earned the name *Rex Pacificator,* 'King-Peacemaker', for his efforts to mediate between England, to which he was allied by his marriage to Henry VIII's sister, and France, Scotland's 'auld ally'. His efforts to prevent war were rejected by Henry VIII, who defiantly declared that he was the 'verie owner' of Scotland. On 22 August 1513, at the head of the finest army Scotland had ever produced, James crossed the Tweed into England.

The Scots met the English army near Flodden Edge on a stormy September afternoon. The Scots made the first breakthrough, but fortunes changed as the bills of the English soldiers proved better weapons than Scottish spears. The Scots refused to give way, and the battle turned into a dreadful massacre. Scotland's ruling class was decimated. Among the dead were the king, nine earls, 14 lords and many chiefs of Highland clans.

James IV died excommunicate, and Henry VIII (who was in France at the time of Flodden) had to get the Pope's permission to have him buried in consecrated ground. There were many who believed that the king was not dead – but had perhaps escaped to France – and would one day return.

James V

Born: Linlithgow Palace, *c.*10 April 1512. **Parents:** James IV and Margaret Tudor. **Ascended the Throne:** 9 September 1513. **Crowned:** Stirling Castle, 21 September 1513. **Reign:** 29 years. **Married:** (1) Madeleine, daughter of Francis I of France, (2) Mary, daughter of the Duke of Guise. **Children:** With (2), two sons and one daughter, Mary, future Queen of Scots; nine illegitimate children. **Died:** Falkland Palace, 14 December 1542. **Buried:** Holyrood Abbey.

In the days of the infant king, Scotland was devastated by the results of Flodden and by internal intrigues. Two main factions arose: the pro-English and (eventually) Protestant; and the pro-French and Catholic.

In 1528 James V escaped from the control of the pro-English faction. Taking over the royal government, he had some success in restoring order, even in the north and west. He married a French rather than an English princess (the first soon died but he married another), resulting in war with the aggressive Henry VIII. Defeated at Solway Moss (1542), he died soon after the birth of his hoped-for heir – a girl.

Mary

Born: Linlithgow Palace, 7 Dec 1542. **Parents:** James V and Mary of Guise. **Ascended the Throne:** 14 Dec 1542. **Crowned:** Stirling, 9 September 1543. **Reign:** 25 years. **Married:** (1) Francis II of France (1558), (2) Henry Stewart, Lord Darnley (1565), (3) James Hepburn, Earl of Bothwell (1567). **Children:** With (2), one son, James VI and I. **Died:** Fotheringhay Castle, Northamptonshire, 8 Feb 1587. **Buried:** Peterborough Cathedral; removed to Westminster Abbey in 1612.

In 1544 Henry VIII of England invaded Scotland in an effort to enforce the marriage of the infant Mary to his son Edward. She was sent to France to marry the Dauphin, later Francis II. A Catholic, she returned, in 1561, to Scotland, now predominantly Protestant, after Francis's death a year earlier and married her unruly cousin, Lord Darnley. They fell out, and Darnley was involved in the plot to murder her secretary, David Rizzio. He was himself killed in 1567. Mary scandalized her subjects by promptly marrying her presumed lover and the suspected murderer of Darnley, the raffish Earl of Bothwell. Forced to abdicate, she fled to England, where she was held prisoner by Elizabeth I. A focus for pro-Catholic intrigue, she was executed 19 years later.

James VI

Born: Edinburgh, 19 June 1566. **Parents:** Mary, Queen of Scots, and Lord Darnley. **Ascended the Throne:** 24 July 1567. **Crowned:** Stirling, 29 July 1567.

James VI was proclaimed king at the age of one on the forced abdication of his mother.

He had a wretched childhood while Scotland was ruled by a succession of regents, and became the puppet of different factions. In 1583, escaping from the captivity of the dominant Protestant lords, James asserted his authority and took over the government, though he was unable to control the strife of the fractious nobility, Protestant or Catholic, or to subdue Presbyterianism, which he regarded as a threat to royal government.

He was, however, one of Scotland's most successful monarchs and was a notoriously shrewd political operator. He ruled his country very effectively, and by the end of the 16th century, his control stretched as far as the Highlands.

He maintained good relations with England (in spite of the execution of his mother), and in 1603 he inherited the English Crown on the death of Elizabeth I (*see* **James I of England,** page 179).

The Union of the Crowns

By 1583, when the 17-year-old James VI took over his royal powers in Scotland, it was obvious that Elizabeth I of England would produce no heir, and England's Tudor dynasty would end with her death.

As her successor, James was really the only likely candidate. He was directly descended from Henry VII (founder of the Tudor dynasty) and, vitally important, he was a Protestant.

He could not wait for Elizabeth to die. In private, he complained that she seemed likely to outlive the sun and moon.

James was nevertheless careful to stay on good terms with England and its queen. The execution of his mother, Mary, Queen of Scots, in 1587 enraged her former subjects, but James made only a formal protest.

ONE CROWN, TWO GOVERNMENTS

The union of the Scottish and English Crowns was just that – one king but two kingdoms. Except for sharing a monarch, England and Scotland were to remain separate until the Act of Union in 1707.

Elizabeth refused to discuss the succession, and ignored broad hints from James that she should name him, but all her courtiers favoured James. Her powerful minister, Sir Robert Cecil, kept up a secret correspondence with him, offering advice and encouragement.

Some said that Elizabeth did eventually name James as her heir, on her deathbed, and thus, despite her earlier reluctance, helped to ensure a peaceful succession.

Irish Kings

Ireland was probably settled by Celts soon after 500 BC. The evidence of the Greek explorer Pytheas, who circumnavigated Britain in the late fourth century BC, suggests that the Celts were certainly in possession by that time. There seems to have been a rebellion against them by the original inhabitants in the first century AD.

The **Romans** never conquered Ireland, although they had substantial outposts on the eastern coast. Archaeology has revealed considerable Romanization, deriving mainly, perhaps, from post-Roman Britain.

From an early date, the Irish tribes were ruled by royal dynasties, supported by an aristocratic élite. Thanks to the ancient Irish interest in genealogy, hundreds of names of kings and dynasties are known, but they remain just names, some no doubt legendary.

There may have been about 100 petty kingdoms in Ireland at any given time. The king was essentially a war leader and representative of his people.

Real power belonged to the kings of the provinces, who were constantly engaged in war and dynastic conflicts.

There were five provinces, or kingdoms, the so-called 'Five Fifths', roughly corresponding to their modern descendants: Ulster; Connacht; North Leinster; South Leinster; and Munster. After AD *c.*500, there were seven, Ulster forming three kingdoms, together with Meath (occupying most of North Leinster), Leinster, Munster and Connacht.

King Cormac and Eithne

On the evidence of the old sagas, a concept of Irish unity existed at a very early period, with the Kings of Connacht and Ulster contending for supreme authority. In the third century AD, the Kings of Connacht gained the upper hand, expanding east and north and making Tara their capital. The semi-legendary figure of **Cormac** (reigned *c.*227-66) played a leading role in their expansion.

As Roman authority declined in the fourth century AD, Irish raids on the coast of Britain increased, and led to settlement. Colonies were founded in Wales and – a permanent occupation – in Scotland (by the Scots of Dalriada).

The greatest leader of the House of Conn (from which Connacht took its name) was the semi-legendary **Niall of the Nine Hostages** (reigned *c.*380-405). He and his kin, the Ui ('descendants of') Neill ruled almost all the northern half of Ireland. One group, the northern Ui Neill, controlled western Ulster; another, the southern Ui Neill, were based in Meath. The Dal Cuinn ('race of Conn') provided the High Kings of Ireland until the early eleventh century.

THE VIKINGS

Norse raids began in 795, and in the ninth century Norsemen founded a number of small coastal kingdoms, forming the nucleus of future cities such as Dublin, Wexford and Waterford. Norse rulers used their Irish kingdoms to launch attacks on England and Scotland.

The Norse invasions

Brian Boru killed by Vikings

halted in the reign of the high king **Aed Finnliath** (862-79). He extinguished all the Norse outposts in northern Ireland, but the Norse Kings of Dublin, such as **Olaf the White,** dominated a large part of the country in the ninth century, sometimes preventing the high king from holding his annual assembly at Tara.

The threat of a Norse conquest was ended by **Brian Boru** (941-1014), Ireland's greatest national hero. He became King of Munster in 976 and in 999 gained the support of the Kings of Leinster and Dublin, after defeating them in battle at Glen Mama, near Dublin. In 1002 he became high king, displacing **Mael Sechnaill III,** the last of the Ui Neill high kings.

The dissatisfied Kings of Leinster and Dublin formed a conspiracy with Sigurd, the Norse Earl of Orkney, who landed with a large Norse army at Clontarf, near Dublin, in 1014. Brian was expecting him, and the Norsemen with their Leinster allies were utterly defeated. After the battle, however, a Norse warrior entered Brian's tent and killed him.

Brian had no worthy successor. Mael Sechnaill regained the high kingship until his death in 1022, but thereafter, for 150 years, Ireland's minor kings battled for supremacy, and no high king gained universal recognition or reigned without a rival.

THE NORMANS

In 1166 the unpopular King of Leinster, **Dermot MacMurrough** (reigned 1134-71) was driven out by a coalition of enemies. With his daughter Eva, he sailed to England to seek help. The English king, Henry II, gave permission for him to raise forces, and he soon enlisted ambitious Norman adventurers in South Wales. Chief among them was Richard de Clare, Earl of Pembroke, known as **Strongbow** (died 1176), who was promised Eva in marriage. Dermot returned to Ireland in 1169, soon followed by his Norman allies.

The invaders quickly achieved success, when a force under the last High King of Ireland, **Rory O'Connor** (Ruadri Ua Conchobar, reigned 1166-86, died 1198), was defeated. On

Dermot's death, Strongbow, now married to Eva, declared himself King of Leinster.

Fearing the creation of an independent Norman kingdom in Ireland, Henry II arrived in person in 1171 to assert his authority. He brought a large army but did not need it. In general, the Irish chiefs received him with relief. Rory O'Connor recognized him as overlord in 1176, though he was later deposed. **Cathal O'Connor,** King of Connacht (1201-24), the last native Irish king, resisted the English for some years.

English rule in the Middle Ages was nominal. Ireland continued to be ruled by local nobles and chiefs, who acknowledged the King of England as their overlord but never set eyes on him.

Henry VIII was the first English king to call himself 'King of Ireland'. Thereafter, English interference steadily increased, and was increasingly resented, until Ireland won independence. Twenty-six counties formed the Irish Free State in 1922; 6 northern counties formed Northern Ireland which remained part of the UK. The Free State became a republic and left the Commonwealth in 1947.

Richard II & Four Kings of Ireland

The High Kingship of Ireland

The High Kingship *(Ard Ri)* of Ireland was more an idea than a functional office, symbolizing that the Celtic Irish were one people. The high king had no established powers, and did not receive any revenue from the lesser kings, but he could settle disputes between them, and he could call out the Irish 'host' to fight invaders – or to raid Britain.

The high kings held annual assemblies attended by large numbers, although there is no evidence that laws were passed there.

The 'capital' of the high kings was the hill-fort of Tara, in County Meath, where the remains of extensive earthworks can still be seen, together with the stone of destiny where the kings are said to have been crowned, and the outline of a banqueting hall 232m (760 ft long). Cormac is said to have founded schools of learning at Tara in the third century, and to have built the five great roads that radiated from Tara across Ireland. It was a great religious centre in pre-Christian times, but it ceased to be a royal residence in the sixth century.

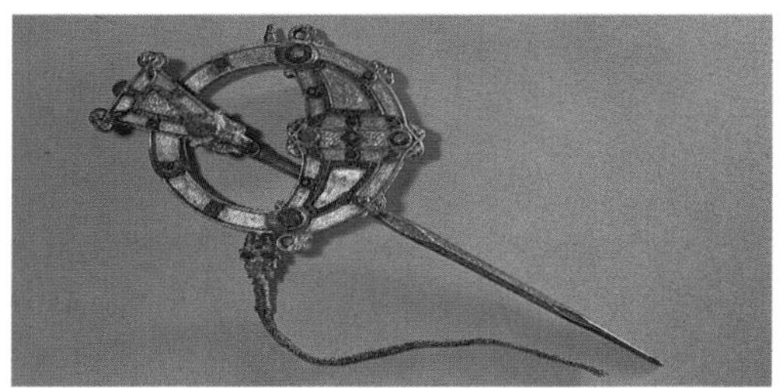

Tara Brooch

Cross at Moone, Co. Kildare

Princes of Wales

The Romans invaded Wales in the first century AD, conquering the Silures (in South Wales) and the Ordovices (in North and Mid-Wales) as well as lesser Celtic tribes. Only South Wales, however, became fully part of the Roman world. Elsewhere, forts were scattered across the country, and many of them were abandoned before the final Roman withdrawal from Britain.

Because Wales is a mountainous country, communications were difficult, presenting many obstacles to invaders but also making it difficult for defenders to assemble – and feed – an army. The kingdoms that developed in the early Middle Ages were therefore small and separate. The necessary preconditions for the formation of a single, nationwide political unit did not exist.

However, as time passed, some of the small kingdoms that emerged after the Roman period proved more successful than others. The most viable were those that commanded stretches of useful lowland, especially Gwynedd in the north (where rugged Snowdonia protected the 'bread-basket' of Anglesey), Dyfed in the south-west and Deheubarth in the south. Powys, in the east, was also a survivor, but suffered by its proximity to England.

Significantly, the great princes of medieval Wales were all westerners, many from Gwynedd. They were able to exercise authority well beyond the borders of their kingdoms and at times claimed to rule all Wales.

These were individual triumphs, which hardly lasted more than one generation, for there was no law of primogeniture to prevent the division of a chief's territory among his heirs. Yet the three dominant kingdoms continued to form the basic political framework until the English conquest.

In 1282 the conqueror of Wales, King Edward I of England, declared his own new-born son 'Prince of Wales', a title still bestowed on the heir to the British Crown. Sporadic resistance to English rule culminated in the rebellion of **Owain Glyndŵr** in 1400. Henry Tudor, whose ancestors included a daughter of Llywelyn the Great, gained the English throne as Henry VII in 1485, and England and Wales were merged in one kingdom by the Act of Union (1536).

Major Welsh Rulers

Cunedda

Lived: *c.* 400.

A British chieftain from, possibly, Lothian, Cunedda and his kin settled in North Wales after driving out the Scots (invaders from Ireland). The Kings of Gwynedd claimed him as their ancestor.

Maelgwyn Hir, 'the Tall'

Authority: King of Gwynedd. **Died:** *c.*547.

Allegedly a grandson of Cunedda, Maelgwyn Hir ruled most of North Wales including Anglesey from his stronghold at Degannwy. Though said to have entered a monastery, he soon returned to society, and gained an unsavoury reputation: he was believed to have murdered his wife and nephew in order to marry the latter's widow.

Cadwallon

Authority: King of Gwynedd. **Father:** Cadfan.
Died: *c.*634.

Driven out by the Northumbrian king, Edwin of Deira, Cadwallon formed an alliance with King Penda of Mercia, regained his kingdom and defeated and killed Edwin at Heathfield, near Doncaster. He failed to make the most of this opportunity to restore British (Celtic) rule and was killed by Edwin's kinsman, Oswald of Bernicia, in a sneak attack near Hadrian's Wall.

Merfyn Frych, 'the Freckled'

Authority: King of Gwynedd. **Reigned:** 825-44.
Married: Nest, daughter of the King of Powys.
Died: 844.

Merfyn Frych inherited Anglesey in 825 and was later recognized as king throughout much of north Wales.

Rhodri Mawr, 'the Great'

Authority: King of Gwynedd. **Reigned:** 844-78.
Parents: Merfyn Frych and Nest.
Married: Angharad, daughter of the King of Ceredigion (Deheubarth). **Died:** 878.

He succeeded his father as King of Gwynedd, his uncle as King of Powys (855) and his father-in-law as King of Ceredigion (Deheubarth), thus uniting most of Wales outside Dyfed and Gwent. Although his dominion did not last, it encouraged the idea of Welsh unity among future generations. Rhodri's reign was prosperous, but he spent much of his life fighting, especially against Viking marauders. He was killed in battle against the Mercians.

Anarawd ap Rhodri

Authority: Prince of Gwynedd. **Reigned:** 878-916.
Parents: Rhodri Mawr and Angharad. **Died:** 916.

Rhodri Mawr's lands were divided on his death, Anarawd receiving part of Gwynedd, including Anglesey. He may have been responsible for a defeat inflicted on the Mercians in 881: he later formed a defensive alliance with the Danish King of York. In campaigns against his brother **Cadell ap Rhodri**, who ruled Ceredigion, he received aid from Alfred of Wessex, whom he acknowledged as overlord. Later rulers of Gwynedd and Deheubarth were descended from Anarawd and Cadell respectively.

Hywel Dda, 'the Good'

Authority: King of Deheubarth. **Reigned:** *c.*904-50.
Father: Cadell ap Rhodri.
Married: Elen, daughter of the King of Dyfed. **Died:** 950.

Hywel Dda inherited Ceredigion (Deheubarth) on his father's death, gained Dyfed by his marriage and, when Idwal Foel (see below) was killed in 942, he took over Gwynedd too. Thus, most of Wales was united, though briefly, under him. He seems to have been a frequent visitor to the court of the House of Wessex and acknowledged the English king as overlord. No mere warlord, he was something of a scholar, who made a pilgrimage to Rome, minted his own silver coinage and compiled a code of law, after first summoning a consultative assembly from all over his territory. His laws contributed to the consciousness of national unity.

Idwal Foel, 'the Bald'

Authority: King of Gwynedd. **Reigned:** 916-42.
Father: Anarawd ap Rhodri. **Died:** 942.

A reluctant vassal, Idwal Foel was killed in a rebellion against the English, and his kingdom passed to his nephew, Hywel Dda (see above).

◀ *Hywel Dda, 'the Good'*

Iago ap Idwal

Authority: King of Gwynedd. **Reigned:** 950-79.
Father: Idwal Foel. **Died:** ?979.

Excluded from the kingdom when his father died, he regained it, in conjunction with his brother Ieuaf, on the death of Hywel Dda. Complex dynastic conflict ended with Iago being deposed by Ieuaf's son. Iago was one of the Welsh princes recorded as paying homage to the English king, Edgar, at Chester in 973.

Maredudd ap Owain ap Hywel Dda

Authority: King of Deheubarth. **Reigned:** 986-99. **Died:** 999.

Maredudd inherited Deheubarth from his father and emulated his grandfather in uniting, briefly, north and south Wales.

Llywelyn ap Seisyll

Authority: King of Deheubarth and Gwynedd. **Reigned:** 1018-23.
Married: Angharad, daughter of Maredudd ap Owain.

He had claims to Gwynedd and, through his marriage, Deheubarth. His warlike abilities made him master of both and, though his reign was short, he laid the basis for the claims of his more famous son.

Gruffydd ap Llywelyn

Authority: King of Gwynedd and Powys. **Reigned:** 1039-63.
Parents: Llywelyn ap Seisyll and Angharad.
Married: Ealgith, daughter of Aelfgar of Mercia.

On succeeding his father, Gruffydd ap Llywelyn gained immediate fame by defeating the Mercians. He then turned against Deheubarth, whose king was defeated and killed. Gruffydd did not, however, gain complete control of the southern kingdom until 1055. He raided the English borders and formed an alliance with King Aelfgar of Mercia. In 1062, he was suddenly attacked by Earl Harold Godwinson of Wessex and killed by traitors.

Gruffydd ap Cynan

Authority: King of Gwynedd. **Reigned:** 1081-1137.
Born: *c.*1055. **Died:** 1137. **Parents:** Cynan ap Iago and Rhagnell, daughter of the Norse King of Dublin.
Married: Angharad, daughter of a Gwynedd chieftain.

Gruffydd ap Cynan was born in Ireland of the royal line of Gwynedd and made several attempts to regain Gwynedd before finally succeeding. Much of his kingdom was overrun by Normans, who imprisoned him but he escaped to join the anti-Norman rebellion of 1094. Driven out again in 1098, he retired to Ireland, but returned as ruler of Anglesey, swearing fealty to Henry I.

Rhys ap Tewdwr

Authority: King of Deheubarth. **Reigned:**1081-1093.
Married: Gwladus, daughter of the King of Powys.
Died: 1093.

A descendant of Hywel Dda, Rhys ap Tewdwr made good his claim to the kingdom at the Battle of Mynydd Carn (1081) with the help of Gruffydd ap Cynan, and was confirmed in it by King William I of England, apparently paying £40 a year. He fought against various rival princes and rebellious kinsmen and, after William's death, was killed in battle against the Normans who were overrunning south Wales.

Madog ap Maredudd

Authority: King of Powys. **Reigned:** 1132-60.
Father: Maredudd ap Bleddyn ap Cynfin.
Married: Susanna, daughter of the King of Gwynedd.
Died: 1160.

The last king of all Powys, Madog was forced to defend it against a prince of Gwynedd but maintained good relations with the Norman Earl of Chester and King Henry II of England. After his death, Madog's kingdom was divided up and never reunited

Owain Gwynedd

Authority: King of Gwynedd. **Born:** *c.*1100. **Reigned:** 1137-70. **Parents:** Gruffydd ap Cynan and Angharad. **Married:** (1) Gwladus, daughter of Llywarch ap Trahaern, King of Gwynedd, (2) Christina, a cousin. **Children:** Two sons by each wife; numerous other children. **Died:** 1170. **Buried:** Bangor Cathedral.

The name Owain Gwynedd serves to prevent confusion with another Owain ap Gruffydd. With his brother he restored Gwynedd during his father's old age. He extended the kingdom and conducted fruitful operations in the south, benefiting from the anarchy in England. But when the English king, Henry II, appeared in North Wales in 1157, Owain Gwynedd recognized the need for prudence, swore fealty, and changed his title from king to prince. He joined the general rebellion against Henry in 1165, when his authority extended as far as the Dee.

Rhys ap Gruffydd, 'The Lord Rhys'

Authority: King of Deheubarth. **Born:** *c.*1133.
Reigned: 1155-97. **Parents:** Gruffydd ap Rhys ap Tewdwr and Gwenlliam, daughter of Gruffydd ap Cynan.
Married: Gwenlan, daughter of Madog ap Maredudd.
Children: Eight sons, one daughter.
Died: 1197 **Buried:** St David's Cathedral.

Rhys ap Gruffydd took part in his first battle at the age of 13. He rendered homage reluctantly to Henry II of England in 1158, giving up some territory and the title of king – hence the name by which history knows him, 'The Lord Rhys'. He recouped his losses after the rebellion of 1165 when the most troublesome Normans were turning their attention to Ireland and Henry was having trouble with the Church. Reconciled with Henry, he remained on good terms and adopted Anglo-Norman customs. His later years were troubled by rebellious sons and Norman neighbours.

Llywelyn ap Iorwerth, 'the Great'

Authority: Prince of Gwynedd.
Born: Dolwyddelan Castle, Gwynedd, 1173.
Reigned: 1194-1240. **Parents:** Iorwerth Drwyndwn, son of Owain Gwynedd, and Margaret, daughter of Madog ap Mareddud. **Married:** Joan, daughter of King John of England. **Children:** One son, four daughters; other illegitimate children. **Died:** 11 April 1240.
Buried: Aberconwy Abbey.

The early years of Llywelyn ap Iorweth were occupied, as was so often the case, with dynastic conflicts. He gradually eliminated his rivals and showed his mettle by capturing Mold from the English (1199). By 1203 he was in undisputed control of all Gwynedd.

Powys remained a potential threat, but within two years he had established his suzerainty there, at the same time seeking to nullify adverse reaction from England by marrying the illegitimate daughter of the king.

Inevitably, friendship with England eventually broke down. He lost some territory in 1211, but was able to recoup as John became embroiled with his barons. Llywelyn co-operated with John's opponents, including the Pope, while bestowing his numerous daughters as wives on the Marcher lords. He extended his operations to South Wales, and was recognized as suzerain by all the Welsh princes.

John signing the Magna Carta containing Llewelyn's rights

In the famous Magna Carta wrung out of King John by the barons in 1215, special clauses were concerned with securing the rights of Llywelyn.

In his later years Llywelyn planned to adopt primogeniture to preserve his princedom and considered a centralized government.

Dafydd ap Llywelyn

Authority: Prince of Wales. **Born:** *c.*1208. **Reigned:** 1240-1246. **Parents:** Llywelyn the Great and Joan. **Married:** Isabella, daughter of William de Breos, a Marcher lord. **Children:** None. **Died:** 25 February 1246. **Buried:** Aberconwy Abbey.

The generally recognized heir of Llywelyn the Great, Dafydd ap Llywelyn took the title 'Prince of Wales' in 1244. Relations with his English overlord were tense, but his position was strengthened when his rival, Gruffydd (an illegitimate half-brother), broke his neck escaping from prison. But Dafydd died young and without an heir, and his dominion was once more divided.

Llywelyn ap Gruffydd, 'Llywelyn the Last'

Authority: Prince of Wales. **Reigned:** 1246-82.
Parents: Gruffydd ap Llywelyn ap Iorwerth and Senena.
Married: Eleanor de Montfort, daughter of Simon de Montfort, 6th Earl of Leicester.
Children: One daughter. **Died:** 11 December 1282.
Buried: Abbey of Cwm Hir.

Llywelyn ap Gruffydd defeated his brothers at Bryn Derwin (1255) and, seizing the opportunity of the barons' revolt (see Henry III), he made himself lord of as much territory as his grandfather, Llywelyn the Great, while forming an alliance with the barons' leader, Simon de Montfort. He was officially recognized as Prince of Wales by the Treaty of Montgomery (1267). The succession of Edward I to the Crown of England brought his ruin. He refused to offer homage to Edward and within a few years his sovereignty was confined to part of western Gwynedd. Renewing his rebellion in 1282, he was killed in a skirmish near Builth.

Owain Glyndŵr

Authority: 'Prince of Wales' (self-proclaimed).
Born: *c.*1354. **Parents:** Gruffydd Fychan ap Madog and Helen, daughter of Thomas ap Llywelyn. **Married:** Margaret Hanmer.
Children: Six sons, several daughters. **Died:** *c.*1416.

A descendant of Madog ap Maredudd and other princes, Owain Glyndŵr led a revolt, originally sparked off by personal grievances, against King Henry IV of England in 1400. By 1405 he controlled most of Wales, holding parliaments and making treaties with foreign powers. He survived the fall of his chief English allies, the Percys of Northumberland, in 1403, but after 1405 his position gradually crumbled, while Henry's improved. He is rarely heard of after 1412, but seems to have lived until 1416 or later.

The Saxons, Normans & Plantagenets

Anglo-Saxon Kingdoms

The Germanic people from northern Germany and Scandinavia who settled lowland Britain in the fifth century AD were essentially tribal groups led by warrior-aristocrats under a chieftain or king. Supposedly descended from the god Wotan, this king was also a law-giver and was sanctified by rites which later developed into what we know as Christian coronation, making the king also the champion of the Church.

The first English historian, the Venerable Bede (Baeda, *c.*673-735), described the Germanic settlers as Angles, Saxons and Jutes. He said that the Angles settled in the east, the Saxons in the south, and the Jutes in Kent. Archaeology suggests that this is roughly correct, although the situation was more complicated than Bede's simple description. Other groups, such as Franks, were certainly also involved.

England was at first divided into numerous little kingdoms, most of whose names and rulers are unknown. The main kingdoms that emerged were: Kent, Essex ('East Saxons'), Sussex ('South Saxons'), East Anglia, Lindsey, Bernicia, Deira, Mercia and Wessex ('West Saxons'). These were soon reduced to seven – the 'Anglo-Saxon Heptarchy'. Lindsey, centred around Lincoln, was absorbed by its neighbours and disappeared. Bernicia and Deira combined to form Northumbria. In fact, the number of major kingdoms varied, and at different periods several of them in turn gained ascendancy over the others. Bede listed seven monarchs who held the title of *'Bretwalda'*, a 'high king' acknowledged as superior by the other kings.

The area settled by the newcomers corresponds roughly to modern England, excluding Cornwall in the far south-west and, in the north, extending beyond the modern border, to the Forth. The British (Celtic) kingdom of Strathclyde included part of Cumbria. Frontiers were fluid, and the invasions of the Vikings caused almost total disruption of the Anglo-Saxon kingdoms in the ninth century.

Little or nothing is known of the early Anglo-Saxon monarchs beyond their names, and most dates are approximate.

Hengist and Horsa established early Anglo-Saxon kingdoms

KENT

Kent was the first of the Anglo-Saxon monarchies, and the first kingdom to achieve a degree of dominance over the others. It was also the first to be converted to Christianity, early in the seventh century. It appears to have maintained cross-Channel links and may have controlled trade with continental Europe. Its capital, Canterbury, became the headquarters of the English Church.

Hengist (died 488) was the founder of the Kentish kingdom (see page 20).

Aesc, or **Oisc** (reigned *c.*488-512) gave his name, 'Oiscings' to the Kentish dynasty.

Octa (reigned *c.*512-40).

Eormenric (reigned *c.*540-60).

Vortigern, the southern high king, granted Hengist land in Kent

Ethelbert I, or **Aethelbert** (reigned: *c.*560-616)

Ethelbert I was acknowledged as *Bretwalda* by most of the other kingdoms. He married a Frankish, Christian princess, Bertha, and was himself converted to Christianity by St Augustine, the missionary appointed to convert the English by Pope Gregory the Great in 597. When he died he was buried in the new abbey church at Canterbury.

Silver penny of Ethelbert

Eadbald (reigned 616-40) also married a Frankish princess, but temporarily abandoned Christianity.

Earconbert (reigned 640-64).

Egbert (reigned 664-73).

Hlothere (reigned 673-85) succeeded his brother, Egbert. He was involved in dynastic contests and conflicts with neighbouring kingdoms, and died fighting the South Saxons on 6 February 685.

Eadric (reigned *c.*685-87) overthrew his predecessor and uncle with South Saxon help.

Oswini (reigned *c.*688-90).

Wihtred (reigned 690-725) was a powerful ruler, who threw off East Saxon influence and issued a code of laws.

Ethelbert II (reigned 725-62, jointly with his brother **Eadbert I** [725-48], half-brother Alric, and Eadbert's son **Eardwulf** [from *c.* 747]) was the father of King Egbert of Wessex.

Egbert II (reigned *c.*765-80).

Ealhmund (reigned *c.*784, when Kent was under Mercian control).

Eadbert II (reigned 796-98) was defeated and deposed by usurpers.

Cuthred (reigned 798-807) had one or two successors but, after the end of Mercian dominance (*c.*825), Kent came under the rule of the Kings of Wessex.

ESSEX

Although probably based on London, Essex, which generally included the land of the Middle Saxons (Middlesex), never achieved much prominence among the Anglo-Saxon kingdoms. At one point Essex kings held some power in Kent, but they were submerged by the powerful Mercians in the eighth century and absorbed by Wessex early in the ninth century.

As in other early Anglo-Saxon kingdoms, many of the listed Kings of Essex were joint rulers.

Aescwine (reigned *c.*527-87) was reputedly the founder of the kingdom.

Sledda (reigned *c.*587-600).

Saebert (reigned *c.*600-16) adopted Christianity.

Sigeberht the Little (reigned *c.*617-50).

Sigeberht the Good (reigned *c.*650-60) restored Christianity after a lapse into paganism.

Swithhelm (reigned *c.*660-65).

Sighere (reigned *c.*665-83, jointly with his uncle Sebbe).

Sebbe (reigned *c.*665-95) was said to have abdicated in order to enter a monastery and was buried at St Paul's, London.

Sigheard (reigned *c.*695-*c.*708)

Swafred (reigned jointly with Sigheard, his brother) visited Rome in 709.

Offa (reigned *c*.709) was Sigheard's son and accompanied Swafred to Rome.

Saelred (reigned 709-46).

Swithred (reigned 746-*c*.758) was a grandson of Sigheard. His capital was at Colchester, where much of the Roman town survived, London being controlled by the Mercians.

Sigeric (reigned 758-98).

Sigered (reigned 798-825) was the last King of Essex.

SUSSEX

Sussex and its kings, like those of Essex, are obscure. The kingdom was said to have been founded by **Aelle,** who landed in 477. Bede named him as the first *Bretwalda,* although at that time there were no other kings of substance to make the title meaningful. His son **Cissa,** after whom the capital, Chichester, was named, succeeded him *c*.514. No other kings are known until the seventh century, when Sussex was dominated by other kingdoms, first by Mercia, and later by Wessex.

Ethelwalh (reigned *c*.685) was converted to Christianity by St Wilfred, who founded the bishopric of Selsey. He was killed by Cadwalla of Wessex (see over).

Berthun (reigned *c.*685), perhaps a co-ruler, was killed in Kent.

Cadwalla (reigned *c.*686-88) was a usurper from Wessex.

Nothelm (reigned *c.*692).

Nunna (reigned *c.*710-25).

Aldwulf (reigned *c.*765).

Osmund (reigned *c.*765-70) was probably the last king before Sussex was absorbed by Mercia.

EAST ANGLIA

The kingdom of the East Angles included what became Norfolk, Suffolk and part of Cambridgeshire. It was relatively self-contained and prosperous. Its kings, most of them shadowy and some unknown, were called 'Wuffings', after the reputed founder of the kingdom. They retained their independence, though latterly under Northumbrian and Mercian dominance, until overrun by the Danes in the ninth century.

Wuffa (reigned *c.*571-78).

Tytila (reigned *c.*578-93).

Raedwald (reigned *c.*593-617) was said to have been a Christian who returned to paganism at the urging of his wife. He was named by Bede as *Bretwalda* of the Anglo-Saxons. He is the most likely person to have been the object of the ship burial at Sutton Hoo near Woodbridge in Suffolk, discovered in 1939. The treasure it contained, some of it from Byzantium, testifies to the wealth of the East Anglian kings.

Eorpwald (reigned *c.*617-27) son of Raedwald, was converted to Christianity by the pressure of King Edwin of Northumbria.

Siegeberht (reigned *c.*631-34), founded the bishopric of

Dunwich, for St Felix, and a monastery. He abdicated in order to become a monk himself.

Siegeberht, the King-Monk

Ecgric (reigned *c*.634-35) was killed resisting King Penda of Mercia.

Anna (reigned *c*.633-54), though apparently installed as Penda's candidate for king, also died fighting the Mercians. He is remembered in legend as the father of several pious daughters.

Ethelhere, or **Aethelhere** (reigned 654), Anna's brother, was killed by Oswy of Northumbria.

Ethelwold, or **Aethelwald** (reigned *c*.654-63) was another of Anna's brothers.

Aldwulf (reigned *c*.663-713), the son of Ethelhere and a Northumbrian princess, also produced saintly daughters who became abbesses.

Aelfwald (reigned 713-49, while East Anglia was under Mercian domination).

Hun Beonna (reigned *c*.749).

Ethelbert (reigned 792), a saint and a martyr, was executed by Offa of Mercia, his father-in-law. He was later buried in St Ethelbert's Cathedral, Hereford.

Athelstan (reigned *c*.828-37) minted coins bearing his image.

Edmund (born *c*.840, reigned *c*.855-70) was the second and last East Anglian king to be canonized, after Ethelbert. He was killed by the Danes on 20 November 870 when he refused to renounce Christianity. Once England's patron saint (later superseded by St George), his remains were ultimately buried at Bury St Edmunds.

King Athelstan

NORTHUMBRIA

The kingdom of Northumbria (the name means 'land north of the Humber') was the first of the Anglo-Saxon kingdoms, after Kent, to gain national ascendancy. At its peak it included all the land between the Humber and the Forth and it was culturally advanced, producing, among other fine treasures, the Lindisfarne Gospels and was home to the first English historian, Bede, a monk at Jarrow.

Northumbria was created from the fusion of Bernicia, centred on Bamburgh and roughly equivalent to modern Northumberland, and Deira, centred on York. They were founded by **Ida** of Bernicia, who reigned *c.*547-59. His son **Aelle** became King of Deira *c.*560-80.

Ethelfrith, or **Aethelfrith,** King of Bernicia (reigned *c.*593-616), was a younger son (possibly grandson) of Ida. He defeated the Picts and the Welsh, capturing Chester, and temporarily conquered Deira.

Edwin, King of Deira (reigned *c.*616-32), was a son of Aelle. He defeated and killed Ethelfrith in alliance with King Raedwald of East Anglia, regaining Deira. He also conquered several smaller kingdoms, including Lindsey, and was acknowledged as *Bretwalda.* He married Ethelburga, daughter of Ethelbert of Kent, and became a Christian under her influence (627). He was killed on 13 October 633, fighting against Penda of Mercia and Cadwallon of Gwynedd, at Hatfield Chase.

Eanfrith, King of Bernicia (reigned 633-34), the son of Ethelfrith, he briefly regained Bernicia before dying in battle.

Oswald (born *c.*605, reigned *c.*634-42) was the son of Ethelfrith and brother of Eanfrith. He reunited Deira and Bernicia and succeeded Edwin as *Bretwalda.* In exile under Edwin, he apparently visited Iona, which inspired him with religious passion. He defeated Cadwallon of Gwynedd in 634, uniting

Bernicia and Deira. In 635 he also asked for a bishop from the Scots, receiving Aidan, to whom he gave the holy isle of Lindisfarne. He was killed in battle against Penda of Mercia.

Oswy (born *c.*602, reigned 641-70), brother or half-brother of Oswald, was the last *Bretwalda* listed by Bede and the last notable Northumbrian king. He threw off Mercian domination and turned the tables by temporarily annexing Mercia and introducing Christianity. In the north he defeated the Britons of Strathclyde, the Scots and the Picts. As his second wife he married Eanfled, daughter of Edwin and Ethelburga. He died on 15 February 670.

Egrith (reigned 670-85) was the son of Oswy and Eanfled. His excursions across the Humber and north of the Forth were repulsed by, respectively, the Mercians and the Picts, who killed him at Nechtansmere, near Forfar.

Aldfrith (reigned 685-704), illegitimate son of Oswy, was a great patron of scholarship and the arts.

Osred I (reigned 704-16).

Coenred (reigned 716-18).

Osric (reigned 718-29).

Ceolwulf (reigned 729-37) was the 'most glorious king', to whom Bede dedicated his *Ecclesiastical History of the English People.* He was deposed in 736, restored, but abdicated in 737 and ended his life as a monk at Lindisfarne.

Eadbert (reigned 737-58) invaded the territory of the Picts and the British of Strathclyde, capturing Dumbarton in 756, but eventually followed his predecessor into a monastery.

Oswulf (reigned 758-59).

Ethelwald Moll (reigned 759-65) defeated a rebellion but was later forced to abdicate.

Alchred (reigned 765-74).

Ethelred I (reigned 774-96 with interruptions) was overthrown in 778, regained his crown in 790, and was killed at Corbridge in 796.

Elfwald I (reigned 778-88) replaced the deposed Ethelred, but was murdered in 788.

Osred II (reigned 788-90) was killed by Ethelred, who then regained the Crown.

Osbald (reigned 796). His reign is said to have lasted less than a month.

Eardwulf (reigned 796-809, with interruptions).

Elfwald II (reigned *c*.808).

Eanred (reigned 809-41) paid homage to Egbert of Wessex in 827.

Ethelred II (reigned 841-63).

Osberht (reigned 850-63, died 866).

Ella, or **Aelle** (reigned 863-67), briefly regained York from the Norsemen in 866, but was killed in battle the following year. His successors were subject to the Norsemen, with restricted authority.

Egbert I (reigned 867-72).

Ricsig (reigned 872-76).

Egbert II (reigned 876-78).

MERCIA

The original kingdom of Mercia occupied roughly what is now the West Midlands. The name refers to 'marches' (border areas), presumably alluding to the proximity of Wales, though the first centre of the kingdom appears to have been on the upper Trent. In Mercia's age of greatness, in the seventh and eighth centuries, its kings ruled all England, except East Anglia, from the Humber to south of the Thames, including London. Its prosperity was largely based on the control of trade, which provided its kings, for virtually the first time, with a reliable and substantial source of revenue.

Creoda (reigned *c.*585-93) is the first documented King of Mercia, supposedly descended from a legendary founder named **Icel.**

King Offa of Mercia

Pybba (reigned *c.*592-606).

Ceorl (reigned *c.*606-26). His daughter married the Northumbrian king, Edwin, who held authority over him.

Penda (reigned 632 (?626)-55) was a renowned warrior, and a pagan, who was involved in a long contest with Northumbria (and was therefore roughly treated by Bede). He defeated the men of Wessex at Cirencester in 628, extending his kingdom south of the Severn. In alliance with Cadwallon of Gwynedd he defeated and killed Edwin of Northumbria in 633, and his successor, Oswald, in 641, attacking, but failing to capture, the Bernician stronghold of Bamburgh, and temporarily ending Mercian subservience to Northumbria. He was finally killed by Oswy in 654 or 655. His conquests were divided among his sons, who became Christians, under Northumbrian suzerainty.

Wulfhere (reigned 657-75) was a younger son of Penda. He threw off Northumbrian supremacy, and invaded Wessex, capturing the Isle of Wight, but lost Lindsey to Egfrith of Northumbria.

Ethelred, or **Aethelred** (reigned 675-704), another son of Penda, regained Lindsey before retiring to a monastery.

Coenred (reigned 704-9), son of Wulfhere, also forsook the rough life of kingship and made a pilgrimage to Rome.

Coelred (reigned 709-16).

Ethelbald, or **Aethelbald** (reigned 716-57) was the scion of a distant branch of the royal house. During his long reign he restored Mercian domination of middle England, and added territory conquered from Wessex. He was murdered by his own people at Seckington, Warwickshire, and was buried at Repton Abbey.

Beonred (reigned 757).

Offa (reigned 757-96) was descended from a brother of Penda. He was the most successful Anglo-Saxon king before Alfred. He reduced the authority of sub-kings and controlled almost all England south of the Humber. He built the earthworks known as Offa's Dyke (still visible) on the western frontier, to exclude raids by the Welsh and facilitate his own raids into Wales. He established a business relationship with Charlemagne and issued an elaborate silver coinage. Some coins bear the head of his queen, Cynethryth, a practice of the Roman emperors with whom he perhaps identified. He began the practice of annointing his son as king in his own lifetime, to end succession conflicts.

Offa, founder of St Albans Monastery

Egfrith (reigned 796), Offa's son, was co-king from 787. He outlived his father by only five months.

Coenwulf (reigned 796-821) belonged to a different branch of the royal house from that of Offa.

Ceolwulf I (reigned 821-23), a brother of Coenwulf, was soon deposed.

Beornwulf (reigned 825) was defeated by Egbert of Wessex and killed by the East Angles. His brief reign marks the end of Mercian ascendancy.

Ludeca (reigned 827).

Wiglaf (reigned 827-40) was deposed by Egbert of Wessex in 829 but subsequently regained the kingship. He was buried at Repton Abbey.

Beorhtwulf, or **Berthwulf** (reigned 840-52).

Burgred (reigned 852-74) seems to have been dependent on Wessex, marrying Ethelswith, daughter of Ethelwulf of Wessex, whom he assisted in his war in north Wales and with whom he co-operated against the Vikings. The Danish host, which was encroaching on Mercia, eventually deposed Burgred, who died in Rome.

Ceolwulf II (reigned 874-*c.*880) was installed by the Danes as king of the unoccupied portion of Mercia, but later deposed.

WESSEX

Founded *c.*494, Wessex originally occupied the southern, central counties of England. It replaced Mercia as the dominant Anglo-Saxon kingdom in the early ninth century, a time when Danish raids were increasing. Wessex alone survived the Scandinavian invasions. It gradually absorbed the other kingdoms, and the Kings of Wessex thus became Kings of England.

Cerdic (reigned 519-34), the first documented King of Wessex (and ancestor of the present queen), landed in Hampshire *c.*494, defeated the local British king and later added the Isle of Wight to his conquests.

Cynric (reigned 534-60), son of Cerdic, expanded the kingdom into Wiltshire.

Ceawlin (reigned 560-91 [?592]), son of Cynric, continued the expansion, his authority extending north of the Thames valley. He held temporary ascendancy over all the southern Anglo-Saxon kings. He was listed by Bede as the second *Bretwalda,* though he was eventually driven from the throne, dying soon afterwards.

Ceorl (reigned 591-97).

Ceolwulf (reigned 597-611).

Cynegils (reigned 611-43) was the first Christian King of Wessex. An attempt to murder King Edwin of Northumbria provoked invasion, and there were also clashes with Penda of Mercia, leading to the loss of the Severn valley sub-kingdom of Hwicce.

Cenwalh (reigned 643-72, with interruptions), son of Cynegils, married and then discarded Penda's sister. The Mercian revenge resulted in loss of territory, including much of Somerset, Hampshire and the Isle of Wight. Cenwalh fled to East Anglia, but subsequently regained his kingdom and founded the cathedral at Winchester (648), where he was buried.

Seaxburgh (reigned *c.*673), widow of Cenwalh, apparently reigned briefly in her own right, the only known Anglo-Saxon queen regnant.

Cenfus (reigned *c.*674).

Aescwine (reigned 674-76).

Centwine (reigned 676-85).

Cadwalla, or **Ceadwalla** (reigned 685-88) began the restoration of Wessex, retaking the Isle of Wight, before abdicating and going to Rome, where he died on 20 April 688.

Ine (reigned 688-726) completed the reconstruction of the kingdom, greatly extending its frontiers to the west, defeating the Britons of Cornwall, as well as the King of Sussex. He was credited with founding the port of Southampton before he abdicated and died on his way to Rome.

Ethelheard, or **Aethelheard** (reigned 726-40) ruled during a period of Mercian predominance.

Cuthred (reigned 740-56).

Sigeberht (reigned 756-57) fell out with his nobles, after killing one of their fellows, and was expelled by his successor, though he held on to Hampshire for a time, before being murdered.

Cynewulf (reigned 757-86) was murdered by Sigeberht's brother.

Berhtric (reigned 786-802) married the daughter of his dominant neighbour, Offa of Mercia. Viking raids on Wessex began during Berhtric's reign.

Egbert (reigned 802-39) was the first of a line of powerful Wessex kings. He spent some early years in exile at the court of

Egbert

Charlemagne and was sub-king of Kent before succeeding his cousin, Berhtric. He married Redburga, a Frankish princess. His defeat of Beornwulf of Mercia at the Battle of Ellandune, near Swindon, in 825 marks the replacement of Mercia by Wessex as the leading kingdom. By the end of his reign, Egbert dominated all England south of the Humber. The lesser kingdoms were finally extinguished, the Cornishmen were defeated, and Danish raiders repulsed. Egbert issued a silver coinage, the earliest known from Wessex.

Ethelwulf (reigned 839-58) was crowned at Kingston-upon-Thames. Son of Egbert, he ruled jointly with his father and his younger brother and, reluctantly, his son Ethelbald. He married Osburga, and four of their sons, including Alfred, became Kings of Wessex. Ethelwulf was involved in incessant wars against the Danes, winning a notable victory at Ockley in Surrey (851). His gold ring is now preserved in the British Museum.

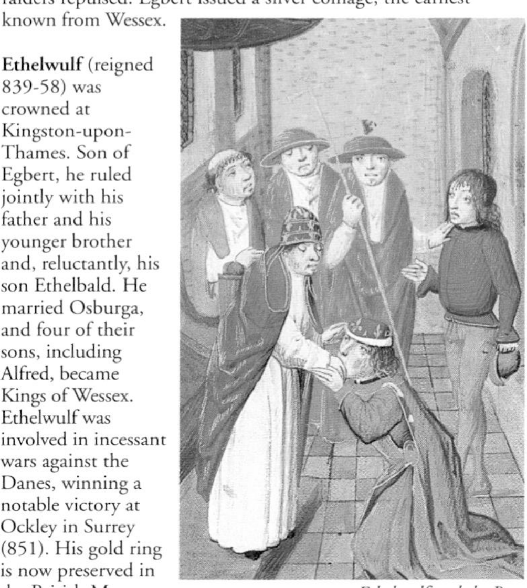

Ethelwulf and the Pope

Ethelbald (reigned 858-60), son of Ethelwulf, married his father's widow, Judith, a Frankish princess, but the marriage was annulled. There were no children.

Ethelbert, or **Aethelbert** (reigned 860-65), previously sub-king of Kent, was unable to prevent the Danes sacking Winchester in 860. He was buried at Sherborne Abbey.

Ethelred, or **Aethelred** (reigned 865-71) was ruling at the time of the full-scale invasion by the Danes' 'Great Army'. In 871, when the all-conquering Danes turned against Wessex, Ethelred, with his younger brother Alfred, fought a succession of battles, receiving mortal wounds at Merton, Oxfordshire. He was buried at Wimborne Minster.

Ethelbald ▶

The Vikings and the House of Wessex

The first Viking raid was recorded by the *Anglo-Saxon Chronicle* in 789. In the next few years northern coasts were raided and the great religious centres of Lindisfarne and Jarrow were sacked. By the 830s the raids of the pagan Vikings – in England mostly Danes – occurred almost every year.

In the 850s the raids became more substantial. The 'Great Army' that invaded East Anglia under Halfdan and Ivarr the Boneless in 865 was bent on conquest and settlement. Within three years it had conquered northern and eastern England. By 874 only one English kingdom remained independent: Wessex.

By a mixture of good luck and good leadership, Alfred of Wessex checked the Danes' advance and preserved Wessex's independence. His successors gradually drove the Danes out and, in the process, became kings of a single country (though it was not yet called 'England').

The Danish leader, Guthrum, made peace with Alfred after Alfred's victory at Edington in May 878. He was baptized a Christian, a symbol of political agreement, and forced to leave Wessex.

By the agreement of 878, made at Wedmore, Somerset, the Danes kept to the region north-west of a line from London to Chester, which became known as the 'Danelaw'.

Viking sea raiders

Saxon Kings of England

Alfred, 'the Great'

Born: Wantage, Berkshire, *c.*849. **Parents:** Ethelwulf of Wessex and Osburga. **Ascended the Throne:** 23 April 871. **Coronation:** Possibly Kingston-upon-Thames, 871. **Married:** Ethelswitha, a Mercian princess. **Children:** Three daughters and three sons, including Ethelfleda, or Aethelfleda, of Mercia and King Edward the Elder. **Died:** *c.*26 October 899. **Buried:** Newminster Abbey, Winchester.

After fierce fighting in 871, the Danes temporarily withdrew from Wessex, but returned in force in 878. Alfred fled to the Somerset levels, then undrained. He emerged from hiding to defeat the Danes at Edington in Wiltshire (May 878). Alfred laid the basis for the unification of England, building a system of fortified towns in Wessex and creating a substantial navy. He devised new laws, based on custom and consultation, and encouraged cultural developments, personally translating several books from Latin.

The chief account of Alfred is the biography written by a monk, Asser, supposedly his contemporary. Recent research has, however, cast doubt on its authenticity. A gold and enamel jewel bearing the words 'Alfred had me made' was found near the Isle of Athelney, Somerset, in 1693. It is now preserved in the Ashmolean Museum, Oxford.

Edward, 'the Elder'

Born: *c.*871. **Parents:** King Alfred and Ethelswitha. **Ascended the Throne:** *c.*26 Oct 899. **Coronation:** Kingston-upon-Thames, 31 May/8 June 900. **Married:** (1) Egwina, (2) Elfleda, daughter of a nobleman, (3) Edgiva, daughter of a Kentish noble. **Children:** With (1), two sons including King Athelstan and (St) Edith; with (2), two sons and eight daughters; with (3), two sons, Edmund I and Edred, and two daughters. **Died:** Farndon-on-Dee, 17 July 924. **Buried:** Winchester Cathedral.

Edward inherited his father's authority over most of England, captured the Danish 'Five Boroughs' (now the East Midlands and Lincolnshire), and built more forts in the sub-kingdom of Mercia, which he inherited on the death of his sister Ethelflaed in 918. Several of his daughters married continental rulers.

SAXON KINGS OF ENGLAND

England became united under the royal house of Wessex during the ninth and tenth centuries. Several monarchs played a part in this, but Athelstan, son of Edward the Elder and grandson of Alfred the Great, is generally regarded as the first king of a whole, united England.

Athelstan

Born: *c.*895. **Parents:** Edward the Elder and Egwina. **Ascended the Throne:** 17 July 924. **Coronation:** Kingston-upon-Thames, 4 September 925. **Authority:** King of England. **Married:** Unmarried. **Died:** Gloucester, 27 October 939. **Buried:** Malmesbury Abbey.

The crucial date for the ascendancy of Athelstan is perhaps 937, when he defeated the Scots and Danes at the Battle of Brunanburh. He was recognized as overlord in Cornwall, Scotland and Wales, although that meant little in practical terms. Like his grandfather, he was remembered also as a law-giver and tactful governor.

Athelstan maintained extensive European connections through the marriages of his numerous sisters.

Malmesbury Abbey ▶

Edmund I, 'the Magnificent'

Born: *c.*921. **Parents:** Edward the Elder and Edgiva of Kent. **Ascended the Throne:** 27 October 939. **Coronation:** Kingston-upon-Thames, 29 November 939. **Authority:** King of England. **Married:** (1) (St) Elgiva, (2) Ethelfleda, daughter of the Ealdorman of Wiltshire. **Children:** With (1), two sons, the future Kings Edwy and Edgar, and one daughter. **Died:** Pucklechurch, Gloucestershire, 26 May 946. **Buried:** Glastonbury Abbey.

Another formidable warrior, Edmund I fought with his predecessor and half-brother, Edward, at Brunanburh. As king he retrieved part of Northumbria conquered by the Norse King of York, Olaf Guthfrithson, and defeated the Britons of Strathclyde, which he turned over to the King of Scots. By 944 his authority was acknowledged throughout England. Edmund was fatally stabbed in a fracas attempting to arrest an outlaw named Liofa.

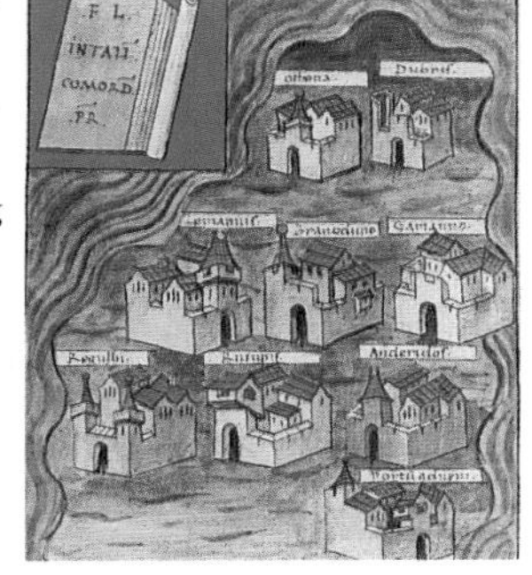

Edred

Born: *c.*923. **Parents:** Edward the Elder and Edgiva of Kent. **Ascended the Throne:** 26 May 946. **Coronation:** Kingston-upon-Thames, 16 August 946. **Authority:** King of England. **Married:** Unmarried. **Died:** Frome, Somerset, 23 November 955. **Buried:** Winchester Cathedral.

In spite of health problems, Edred proved an effective war leader, expelling the Norsemen who had reclaimed York by defeating their leader, Eric Bloodaxe (died 954), and re-enforcing royal authority in Northumbria.

Edwy, 'the Fair'

Born: *c.*941. **Parents:** Edmund I and (St) Elgiva. **Ascended the Throne:** 23 November 955. **Coronation:** Kingston-upon-Thames, *c.*26 January 956. **Authority:** King of England. **Married:** Elgiva (Elgifu), his stepsister. **Died:** Gloucester, 1 October 959. **Buried:** Winchester Cathedralè.

Edwy antagonized the Archbishop of Canterbury, Oda, and the influential Abbot of Glastonbury, (St) Dunstan, by his secret marriage to his stepmother's daughter. According to the Anglo-Saxon Chronicle, the marriage was later annulled. Elgiva died in Gloucester in suspicious circumstances in September 959, and the king died soon afterwards.

Edgar

Born: *c*.943. **Parents:** Edmund I and (St) Elgiva. **Ascended the Throne:** 1 October 959. **Coronation:** Bath Abbey, 11 May 973. **Authority:** 'King of the English and of the other people living within Britain'. **Married:** (1) Ethelfleda, daughter of an ealdorman, (2) Elfrida, daughter of the ealdorman of Devon. **Children:** With (1), one son, the future King Edward the Martyr; with (2), two sons, Edmund and Ethelred II; one illegitimate daughter, (St) Edith, or Eadgyth. **Died:** Winchester, 8 July 975. **Buried:** Glastonbury Abbey.

Edgar was sub-king of Mercia and Northumbria from 957 before succeeding as King of England. He favoured (St) Dunstan who, as Archbishop of Canterbury (961-88), had a powerful influence on Edgar during his largely peaceful reign. He was acknowledged as overlord by the King of Scots, to whom he surrendered Lothian.

King Edgar

Edgar's late coronation followed rites devised by (St) Dunstan, which form the basis of the coronation ceremony today (previous kings had been consecrated rather than crowned). Elfrida was crowned with him, and this was another innovation.

(St) Edward II, the Martyr

Born: *c.* 963. **Parents:** King Edgar and Ethelfleda. **Ascended the Throne:** 8 July 975. **Coronation:** Kingston-upon-Thames, 975. **Authority:** King of England. **Married:** Unmarried. **Died:** Corfe Castle, Dorset, 18 March 978. **Buried:** Wareham, Dorset; in 980 translated to Shaftesbury Abbey.

Edward, though influenced by (St) Dunstan, failed to prevent an anti-monastic reaction during his reign.

His stepmother, Elfrida of Devon, had some support for her belief that her own son, Ethelred, ought to be king. On a visit to them at Corfe, the young King Edward was ambushed and murdered. (The final resting place of his alleged remains, unearthed by archaeologists in 1931, caused a dispute. At one time they were in the custody of the Midland Bank, Croydon.)

Edward being murdered at Corfe Castle

Ethelred II, 'the Unready'

Born: *c.*968. **Parents:** King Edgar and Elfrida.
Ascended the Throne: 18 March 978.
Coronation: Kingston-upon-Thames, April 978.
Authority: King of England. **Married:** (1) Elfleda, or Elgiva, daughter of an ealdorman, (2) Emma, daughter of the Duke of Normandy. **Children:** With (1), eight sons, including Edmund II, and five daughters; with (2), two sons, including Edward the Confessor, and one daughter. **Died:** London, 23 April 1016. **Buried:** St Paul's Cathedral.

Called *'unraed'*, or 'the Redeless', meaning 'uncounselled' rather than 'unready', Ethelred failed the first test of medieval kingship – success as a war leader. He could not prevent new Viking attacks and was forced to buy off King Olaf Tryggvesson of Norway and King Sweyn of Denmark when, provoked by Ethelred's massacre of Anglo-Danes, they sailed up the Thames to threaten London in 994. Famine and an epidemic among the cattle exacerbated his woes. In 1013 Sweyn forced Ethelred from the throne and into exile in Normandy. Ethelred returned, however the following year, after Sweyn's early death, to rule a short while longer.

Danish Kings

Between 875 and 954 a number of Danish and Norse kings preserved a semi-independent enclave in the Scandinavian kingdom of York. From 994 a more formidable Danish assault was launched against England, resulting in a brief dynasty of Danish kings.

Danish raiders attacking the English Coast.

Sweyn, 'Forkbeard'

Born: unknown. **Parents:** King Harold Bluetooth of Denmark; mother uncertain. **Ascended the Throne:** Denmark 986, England 1013. **Coronation:** Not crowned in England. **Authority:** King of Denmark, Norway and England. **Married:** (1) Gunhilda, probably a Polish princess, (2) Sigrid, former wife of the King of Sweden. **Children:** With (1), two sons, including Canute; with (2), one daughter; four other daughters with either (1) or (2). **Died:** Gainsborough, Lincolnshire, 3 February 1014. **Buried:** London; later removed to Roskilde Cathedral, Denmark.

A formidable warrior, Sweyn seized the Danish Crown from his father and created an empire centred on the North Sea. After several earlier attacks, he drove out Ethelred in the autumn of 1013 and was accepted as king. A few weeks later, however, he died as a result of a fall from his horse.

Sweyn dreamt of his own death by drowning

Edmund II, 'Ironside'

Born: *c.*990. **Parents:** Ethelred II and Elfleda. **Ascended the Throne:** 23 April 1016. **Coronation:** St Paul's Cathedral, London, *c.*25 April 1016. **Authority:** King of England. **Married:** Edith (or Eagwyth), widow of an East Anglian thegn. **Children:** Two sons, including Edward the Atheling. **Died:** London (or possibly Oxford), 30 November 1016. **Buried:** Glastonbury Abbey.

Edmund grew up during a period when England's fortunes were at a low ebb because of repeated Viking incursions. Perceptive even in his youth, Edmund recognised the flaws in his father's policy of buying off the Vikings, and as a prince encouraged the country to stand up against them. His valour earned him the epithet 'Ironside'.

Even before the death of his father, Edmund had made himself ruler in Danelaw, However, his efforts to oppose the invasion of Wessex by his rival, Canute, in late 1015 were undermined by the treachery of Ealdorman Edric of Mercia, and in the following year Edmund was unable to hold Northumbria against Canute.

When Ethelred died in 1016, London and the Witan members there chose Edmund as king, but the Witan in Southampton opted for Canute. At Olney, Edmund and Canute agreed to partition England, though the longer lived would succeed to the whole. However, a few weeks later Edmund died after a reign of only seven months and his infant sons fled Canute's rule to settle in Hungary.

Canute, or Cnut

Born: *c.*995. **Parents:** King Sweyn and Gunhilda of Poland. **Ascended the Throne:** 30 November 1016. **Coronation:** Possibly St Paul's Cathedral, London, *c.*1017. **Authority:** King of England, Denmark (from 1019) and Norway (from 1028). **Married:** (1) Elgiva, daughter of the Ealdorman of Deira, never consecrated; (2) Emma of Normandy, widow of Ethelred II. **Children:** With (1), two sons, Harold Harefoot and Sweyn; with (2), one son, Harthacanute, and two daughters; several illegitimate children. **Died:** Shaftesbury, 12 November 1035. **Buried:** Winchester Cathedral.

Canute succeeded to the whole English kingdom on the death of Edmund Ironside. He spent most of his time in England though Denmark occupied him increasingly from the 1020s. While ruthless in establishing his position, and maintaining a Danish standing army, he afterwards earned respect by pursuing statesmanlike policies, founding monasteries and encouraging trade. He employed Englishmen as well as Danes.

Legend relates how Canute mocked his courtiers' obsequiousness by disproving their assertion that the tide would retreat at his command.

Harold I, 'Harefoot'

Born: *c.*1016. **Parents:** King Canute and Elgiva.
Ascended the Throne: 12 November 1035.
Coronation: Oxford, 1037. **Authority:** King of England.
Married: Elgiva. **Children:** One son.
Died: Oxford, 17 March 1040. **Buried:** Possibly Westminster Abbey; disinterred by his successor.

Harold I ruled England as regent for his younger half-brother, Harthacanute, King of Denmark, but in 1037 claimed the Crown in his own right. This caused a breach with the queen dowager, Emma, who went into exile in Flanders.

Harold I

Harthacanute (or Hardicanute)

Born: *c.*1018. **Parents:** King Canute and Emma of Normandy. **Ascended the Throne:** 12 November 1035. **Coronation:** Possibly Canterbury Cathedral, June 1040. **Authority:** King of Denmark and England. **Married:** Unmarried. **Died:** Lambeth, London, 8 June 1042. **Buried:** Winchester Cathedral.

Harthacanute's reign was brief but violent. He arrived with a large fleet to claim the Crown on the death of Harold Harefoot, whose body he had thrown into a marsh. Excessive taxes made him unpopular. He burned Worcester after the murder of royal tax-collectors. He died of 'a horrible convulsion' while attending a wedding.

Queen Emma, mother of Harthacanute.

Edward, 'the Confessor'

Born: Islip, *c.*1004. **Parents:** Ethelred II and Emma of Normandy. **Ascended the Throne:** 8 June 1042. **Coronation:** Winchester, 3 April 1043. **Authority:** King of England. **Married:** Edith, daughter of Earl Godwine of Wessex. **Died:** Westminster, 5 January 1066. **Buried:** Westminster Abbey.

Although Edward was half-brother to Harthacanute, his accession restored the royal house of Wessex to the English throne. He was, though, more Norman than English, having spent the 28 years of Danish rule at the Norman court. Norman influence in England expanded greatly during his reign. It was resisted by the English, especially the powerful Earl Godwine (died 1053) and his son Harold, virtual ruler in Edward's later years.

Pious to a fault, Edward was canonized a century after his death. On religious grounds, he declined to consummate his marriage, and lack of a direct heir promised conflict over the succession. Edward allegedly promised the Crown to Duke William of Normandy, but on his deathbed he was said to have named Harold Godwineson as his successor.

Harold II

Born: *c.*1020. **Parents:** Earl Godwine of Wessex and Gytha, a Danish princess. **Ascended the Throne:** 5 Jan 1066. **Coronation:** Westminster Abbey, 6 Jan 1066. **Authority:** King of England. **Married:** Ealgith, daughter of the Earl of Mercia and widow of Gruffydd ap Llywelyn. **Children:** One or two sons; six or seven illegitimate children with Edith 'Swan-neck'. **Died:** Senlac, Sussex, 14 October 1066. **Buried:** Pevensey Bay, later allegedly removed to Waltham Abbey by Edith Swan-neck.

The man with the best hereditary claim to the Crown upon Edward the Confessor's death was Edgar the Atheling, grandson of Edmund Ironside (and brother-

in-law of Malcolm III, King of Scots). He stood little chance, however. Harold was the man in possession, he had a proven record, and he was unanimously approved by the Witan.

Other claimants proved more formidable. They included King Harold Hardrada of Norway, supported by King Harold's brother, Tostig. They invaded Northumbria but were defeated and killed by Harold at Stamford Bridge near York on 25 September 1066.

Three days later another claimant, Duke William of Normandy, landed in Sussex. He maintained that Harold had recognized him as heir to the Crown when shipwrecked in Normandy in 1064. Marching rapidly south, Harold confronted him near Hastings, and was killed in battle.

Death of Harold

The Normans

Descended from Viking settlers, the Normans were the most formidable people in eleventh-century Europe. They founded a kingdom in Sicily as well as in England. The Duke of Normandy was a more substantial figure than his nominal overlord, the King of France.

The Norman Conquest brought great changes to England, although recent historians point out, first, that this was a period of great change all over Europe and, second, that surviving historical documents are far more numerous for this period than for before, which may make long-established arrangements appear as new.

There were great cultural changes. The Normans made French, rather than English, the everyday language of the élite.

The cultural developments most obvious today were architectural. The Normans were great builders: many of their castles and churches still stand. The style, known as Romanesque on the continent, is in Britain called Norman.

The Normans were well-organized. It is now generally believed that they did not introduce feudalism, in which men held land in exchange for service, but, rather, made it more pervasive and coherent.

The Norman kings were continental rulers. Accordingly, England was drawn closely into European affairs. From 1066 until the fifteenth century, the King of England was also a French prince (for part of that time he claimed to be King of France). European dynastic conflicts became a dominant concern of government.

The Norman Conquest

King Harold had hoped to take the invading Normans by surprise, at the cost of fielding a small and exhausted army, which had marched 240 miles in 13 days. The English forces, holding a hill-top, were well placed, but a Norman feint lured them from their position and they were defeated. The Normans advanced on London, ravaging Kent and burning Southwark. Opposition faded, and William was accepted as king, but five years passed, with frequent rebellions, before Norman rule was secure.

The Normans entrenched their rule by building castles to control towns and other strategic points. Gradually, the big English landholders were replaced by Normans. Eventually, almost the whole ruling class, not merely the dynasty, was changed.

DEATH OF HAROLD

The Norman Conquest is recorded in the 72 scenes of the Bayeux Tapestry. The scene of Harold's death seems to show that he was killed by an arrow in the eye. Closer examination shows that this is a different figure. Harold was killed by a sword.

WILLIAM'S CORONATION

It was said that cheers in Westminster Abbey as William was crowned made the guards outside think a riot had broken out, and they set fire to the surrounding buildings. The consecration went on inside in spite of the flames, though even William, legend has it, was trembling like a leaf.

The Coronation of William I

William I, 'the Conqueror'

Born: Falaise, 1027 or 1028. **Parents:** Robert, Duke of Normandy and Herleva, a tanner's daughter.
Ascended the Throne: 25 December 1066.
Coronation: Westminster Abbey, 25 December 1066.
Authority: King of England and Duke of Normandy.
Married: Matilda, daughter of the Count of Flanders, *c.*1050.
Children: Four sons, including the future kings, William II and Henry I, and six daughters. **Died:** 9 September 1087.
Buried: St Stephen's Abbey, Caen, Normandy.

Known as William the Bastard, he succeeded his father, who had no legitimate heir, as Duke of Normandy in 1035. He became Count of Maine by conquest in 1063. He claimed the English Crown, formally, by right of inheritance: he was first cousin once removed to Edward the Confessor, and his wife was directly descended from Alfred the Great. More realistically, he also claimed it by conquest.

Willliam receives allegiance

William suppressed a widespread rebellion in the north in person (1069). Afterwards he ravaged the country so severely it took generations to recover.

♛ William ordered a detailed inventory of his new kingdom, Domesday Book (1086), an example of Norman efficiency.
♛ In later years, William is reputed to have grown very fat.
♛ William died, campaigning in France, after the pommel of his saddle struck him in the stomach.

William II, 'Rufus'

Born: Normandy, *c.*1056. **Parents:** William I and Matilda of Flanders. **Ascended the Throne:** 9 September 1087. **Coronation:** Westminster Abbey, 26 September 1087. **Authority:** King of England. **Married:** Unmarried. **Died:** New Forest, Hampshire, 2 August 1100. **Buried:** Winchester Cathedral.

William, named 'Rufus' because of his ruddy complexion and red hair, was the third son of William I. The eldest, Robert Curthose, succeeded as Duke of Normandy; the second, Richard, had predeceased his father. Rufus crushed revolts of Anglo-Norman vassals in support of Robert and later gained control of Normandy – which was vital in order to end the dual loyalty of his chief vassals – when Robert went on crusade. He invaded Scotland, killing Malcolm III, and forced obeisance among Scots and Welsh.

DIVINE DISAPPROVAL

Rufus incurred the hostility of the Church, keeping bishoprics and abbeys vacant to milk their revenue and driving the reforming Archbishop of Canterbury, Anselm, into exile. Since chronicles were written by monks, Rufus acquired a poor reputation.

AN EFFETE COURT

The chronicler William of Malmesbury hinted that Rufus was homosexual, surrounded by young men who 'rival young women in delicacy'. Rufus seems never to have shown interest in the latter.

Like his brother Richard (gored by a stag), Rufus died in the New Forest. He was shot while hunting by a knight named Walter Tyrel. Tyrel said it was an accident. There were no other witnesses. The Rufus Stone in the New Forest marks the alleged site of his death.

Henry I

Born: Selby, Yorkshire, *c.*September 1068. **Parents:** William I and Matilda of Flanders. **Ascended the Throne:** 3 Aug 1100. **Coronation:** Westminster Abbey, 5 August 1100. **Authority:** King of England and Duke of Normandy (from 1106). **Married:** (1) Matilda (born Edith), daughter of Malcolm III, King of Scots, and (St) Margaret, (2) Adelicia, or Adelaide, daughter of Geoffrey VII, Count of Louvain. **Children:** With (1), two sons and two daughters; other illegitimate children. **Died:** St Denis le Fermont, near Rouen, 2 December 1135. **Buried:** Reading Abbey.

English-born and well-educated (probably, as a younger son, with an ecclesiastical career in mind), Henry consolidated his position by vigorous efforts to reconcile the English and the Normans. He issued a charter upholding the rights of Englishmen and retracting Rufus's unpopular taxes, and he married a Scottish princess of Anglo-Saxon descent. His chief rival was his brother, Robert of Normandy, who invaded England after returning from the First Crusade in 1101. He was repulsed, and in 1106 Henry conquered Normandy, imprisoning Robert at Cardiff until his death in 1134. Henry then had to defend the duchy against Louis VI of France, who supported Robert's son.

CHURCH AND STATE

To end the conflict between secular and ecclesiastical authority, Henry restored Anselm as Archbishop of

Canterbury. But Anselm refused to do homage to the Crown for church estates and again sought sanctuary abroad. Generally diplomatic, Henry accepted a compromise in 1107.

THE WHITE SHIP

Henry's life was blighted in 1120 when the 'White Ship', bearing both his sons, sank in the Channel. He made his barons swear to accept his daughter, Matilda, as his heir. Henry formed the first English zoo – in the Tower of London. He died of 'a surfeit of lampreys', no doubt food-poisoning.

Henry I on hearing news of his sons' death

Stephen

Born: Blois, *c.*1097. **Parents:** Stephen, Count of Blois, and Adela, daughter of William I. **Ascended the Throne:** 22 Dec 1135. **Coronation:** Westminster Abbey, 26 Dec 1135. **Authority:** King of England. **Married:** Matilda, daughter of Eustace III, Count of Boulogne. **Children:** Three sons and two daughters. **Died:** Dover, 25 Oct 1154. **Buried:** Faversham Abbey.

Stephen was Count of Boulogne through his wife and held large estates on both sides of the Channel. When Henry died suddenly, Stephen was in Boulogne, only a day's journey from London and nearer than other possible claimants. He was accepted by the Londoners and, with most of the English barons behind him, seized the throne, disregarding his oath of fealty to Henry's daughter, Matilda. Stephen could have arrested Matilda when she landed at Arundel in 1141, but he allowed her to join her supporters in Bristol. The resulting civil war lasted 18 years. When Matilda gained the upper hand in April 1141, Stephen was briefly deposed and imprisoned. But Matilda proved even less popular and by December Stephen was back on the throne. He was crowned again, in Canterbury, and a third time at Lincoln in 1146.

Silver penny of Stephen ▶

DIFFICULT DECISIONS

The great Norman barons held lands in England and Normandy. As Matilda's husband, Geoffrey of Anjou, had conquered Normandy by 1144, they were faced with a difficult decision. If loyal to Stephen, they lost their Norman estates, if to Matilda, their English ones.

Matilda, 'the Empress'

Born: London, February 1102.
Parents: Henry I and Matilda of Scotland.
Ascended the Throne: (22 December 1135). **Coronation:** Never crowned. **Married:** (1) Henry V, Holy Roman Emperor, (2) Geoffrey IV, Count of Anjou. **Children:** With (2), three sons, including Henry II. **Died:** Normandy, 10 September 1167. **Buried:** Fontrevault Abbey.

The supporters of Matilda in England were led by her half-brother, Robert of Gloucester. After the Empress, as Matilda was called, arrived in England in 1141, they captured Stephen at Lincoln, and Matilda was proclaimed 'Lady of the English' rather than queen. Stubborn and haughty, she antagonized many people and soon lost London. Stephen regained the initiative and the Crown. In 1148 Matilda retired to Normandy, which her husband, Geoffrey of Anjou, had conquered, and never came back.

STALEMATE

Neither side in the civil war could gain a decisive victory, but the constitutional problem was resolved, after the death of Stephen's son Eustace, by the Treaty of Westminster (1153). Stephen remained king but acknowledged Matilda's son Henry as his heir. (Stephen had another son living, but he did not want to be king.)

♛ A more resolute leader than Stephen was his wife, another Matilda (a favourite Norman name), of Boulogne.

♛ The Empress agreed to release Stephen after his capture at Lincoln in 1141 in exchange for Robert of Gloucester (died 1147), who was held by Stephen's supporters.
♛ Matilda's husband, Geoffrey of Anjou, gave his name to the next royal dynasty, the 'Angevins'.

Geoffrey of Anjou

The Angevins

The Angevins took their name from Geoffrey of Anjou, husband of the Empress Matilda and father of Henry II. From the fifteenth century they were also called Plantaganets. According to tradition, this too derived from Geoffrey. He gained the nickname 'Plantagenet' because he used to wear a sprig of bloom, *planta genista,* in his hat.

Farther back, legend says, the Angevins were descended from a witch, who married the Count of Anjou but disappeared when required to attend Mass.

The Angevins reigned in England until the end of the fifteenth century. Although the succession was not always smooth and rebellion and civil conflict were not infrequent, the power of the Crown was greater than in

The murder of Becket

any other European kingdom, including France. It represented the unity of England, steadily becoming a nation, and there was no doubt that the king was a very special figure. Few people would have contradicted Shakespeare's Richard II: 'There's such divinity doth hedge a king…'. The ceremony of coronation stressed this religious aspect.

The monarch's authority stretched to every part of the kingdom. Landholders and office-holders alike were dependent on royal support and approval.

Royal government was of course changing and becoming more complex. One result was that kings became less peripatetic, tending to settle in or around London, especially at Westminster, the site of the royal shrine (Westminster Abbey), of developing government departments and of meetings of 'parliament'.

Richard I

Henry II

Born: Le Mans, 5 March 1133. **Parents:** Geoffrey of Anjou and the Empress Matilda. **Ascended the Throne:** 25 October 1154. **Coronation:** Westminster Abbey, 19 December 1154. **Authority:** King of England, Duke of Normandy, Duke of Aquitaine, other titles. **Married:** Eleanor, daughter of William X, Duke of Aquitaine, and former wife of King Louis VII of France. **Children:** Four sons, including the future kings Richard I and John, and three daughters; other illegitimate children. **Died:** Chinon, near Tours, 6 July 1189. **Buried:** Fontrevault Abbey.

Henry held vast dominions in France, including, through Queen Eleanor, Aquitaine, and he was the first English king for over a century to inherit the Crown without opposition. His first task was to recover royal authority and restore order and prosperity. Constantly on the move around his empire, he campaigned in the north, regaining land from the Scots, in Wales and in Ireland, where he was accepted as king. He spent most of his reign on the continent, where he further extended his family's estates. His last years were marred by the rebellions of his sons, encouraged by Queen Eleanor and supported by the King of France.

PEACE AND JUSTICE

Intelligent, determined and energetic, Henry is remembered as a great reformer, a founder of English common law, and creator of sound administration and justice. All this is true, for Henry as king was a thorough professional, but, like most medieval kings, he was more interested in dynastic politics than the routines of government and law.

A FLIRTATIOUS QUEEN

Eleanor was 11 years older than Henry, and is said to have seduced him although, as she was a great heiress, that was probably not difficult. Improbable rumour made Saladin a former lover of Eleanor. In any case, Henry was notoriously unfaithful to his wife, and they eventually separated. The Queen died in Normandy in 1204 at a great age, about 82. Henry was worn out after 34 years as king. News that his favourite son, John, had joined a rebellion is said to have hastened his death.

The King and the Archbishop

In 1162 Henry II had his friend and chancellor, Thomas Becket, elected Archbishop of Canterbury. He hoped for an alliance of Church and State, but Becket was determined to be a champion of the Church and became an implacable opponent of the Crown. Henry, with his passion for law and order, wanted criminal priests punished in the royal courts. Becket insisted they should be dealt with by Church courts.

Henry's Constitutions of Clarendon laid down relations between Church and State (among other legal matters). Becket at first agreed, then changed his mind. Henry forced him into exile, but when he returned, five years later, he caused new diff-iculties. In temper, Henry demanded, 'Will no one rid me of this turbulent priest?' Four loyal knights

Thomas Becket ▶

heard him. On 29 December 1170, they rode to Canterbury Cathedral and slew Becket on the altar steps.

The murder of Becket shocked all Europe. He was soon regarded as a saint (canonized 1173). Henry himself, walking barefoot, was an early pilgrim to his tomb.

The shrine of St Thomas was the destination of the pilgrims portrayed by Geoffrey Chaucer (*c.* 1340-1400) in his *Canterbury Tales.*

Richard I

Born: Beaumont Palace, Oxford, 8 September 1157. **Parents:** Henry II and Eleanor of Aquitaine. **Ascended the Throne:** 6 July 1189. **Coronation:** Westminster Abbey, 2/3 September 1189. **Authority:** King of England, Duke of Normandy, Duke of Aquitaine, other titles. **Married:** Berengaria, daughter of Sancho V of Navarre and granddaughter of Alfonso VII of Castile. **Children:** Two illegitimate sons. **Died:** Limousin, 6 April 1199. **Buried:** Fontrevault Abbey.

Richard spent less than a year of his reign in England, though he was, and remains, a popular national hero. When he succeeded, he was allied with Philip Augustus of France in rebellion against his father. As one of the leaders of the Third Crusade, he set out for the Holy Land in 1190, won some famous victories including one against the famous Muslim commander, Saladin but failed to recapture Jerusalem. On his way home he was captured in Austria and held prisoner for 14 months, until ransomed at great cost. During his imprisonment, Philip regained much French territory while, in England, Richard's brother John attempted to displace him. With his remarkable military abilities, Richard regained most of what had been lost before dying of a wound suffered besieging a castle near Limoges.

THE END OF THE ANGEVIN EMPIRE

If Richard, the greatest warrior-king in Europe, had not

died in 1199, the battle between Angevin and Capetian (the French royal house) for dominance in France might have ended differently. Richard is often blamed for neglecting England and draining it of resources, but his first duty was to his dynastic inheritance.

SINS OF SODOM

- Richard's sexual orientation has been much debated. Churchmen tended to preach to him on the fate of the Biblical Sodom, and it was rumoured that he preferred the Queen's brother to the Queen. However, he had at least one, probably two, illegitimate children.
- Richard chose excellent servants, and good government was not neglected under the wise rule of Hubert Walter (died 1205), Archbishop of Canterbury.

John

Born: Beaumont Palace, Oxford, 24 December 1166. **Parents:** Henry II and Eleanor of Aquitaine. **Ascended the Throne:** 6 April 1199. **Coronation:** Westminster Abbey, 27 May 1199. **Authority:** King of England and Ireland, other titles. **Married:** (1) Isabella of Gloucester (annulled 1199), (2) Isabella, daughter of the Count of Angoulême. **Children:** With (2), two sons, including the future Henry III, and three daughters; other illegitimate children. **Died:** Newark, 18 October 1216. **Buried:** Worcester Cathedral.

John was made King of Ireland by his father in 1177. When Richard I died without an heir, the English and Norman barons chose John, other French vassals preferred Arthur of Brittany, his nephew. John disposed of Arthur, perhaps ordering his murder, and regained part of the French possessions, but was formally dispossessed by Philip Augustus in a legal judgment of 1202. In the ensuing wars John eventually lost virtually all of them, including Normandy.

PAPAL INTERDICT

John's refusal to accept Stephen Langton as Archbishop of Canterbury antagonized Pope Innocent III, who laid the kingdom under an interdict (no church services), 1208-13, and excommunicated the King. No one took much notice, but the withdrawal of the Church's approval inspired enemies and rebels.

MAGNA CARTA

John provoked the English barons into revolt, though their economic difficulties through high inflation were not his fault. Civil war broke out, and John was forced to sign the document later known as Magna Carta. It provided redress of grievances, but later ages took it as a statement of civil liberties.

- Traditionally the most unpopular of kings, John has found greater approval among some recent historians.
- John had charm but, spoiled as a child, no tact. Aged 18, he had to be recalled from Ireland after mocking the dress and customs of Irish princes.
- Legend records how, taking a short cut across the Wash in 1216, John lost the crown jewels when caught out by the tide.

King John hunting

The Earl Marshal

William Marshal (*c.*1146-1219) was one of the greatest servants of the Angevin dynasty, and served four monarchs with devoted loyalty. By marrying the daughter of Strongbow he became Earl of Pembroke and he gained his military fame in wars against the French. He supported Henry II against the rebellious Richard who, as king, recognized his worth and made him marshal, an office from which he took his surname. He fought with Richard in Normandy and supported John against the barons, becoming his chief adviser. On John's death he was by common consent appointed regent for the young Henry III.

'Marshal' was an old military title. It corresponded to the French *maréchal* and the Scottish 'marischal'. The marshal was, in effect, the army commander, under the king. In peacetime it became the marshal's business to organize royal journeys and oversee court ceremonies.

In England the office of Earl Marshal became hereditary among William's descendants, passing eventually to the Dukes of Norfolk – in recent centuries the Howards. It lost its military aspect (though the title survived in 'field marshal') and became largely ceremonial. The Earl Marshal's most important job is to organize the coronation.

In Scotland, the hereditary office of Marischal, later Earl Marischal, was held by the Keiths.

Henry III

Born: Winchester, 1 Oct 1207. **Parents:** King John and Isabella of Angoulême. **Ascended the Throne:** 18 Oct 1216. **Coronation:** Gloucester Cathedral, 28 Oct 1216. **Authority:** King of England, Ireland and parts of France. **Married:** Eleanor, daughter of Raymond Berenger IV, Count of Provence. **Children:** Six sons, including the future Edward I, and three daughters. **Died:** Westminster, 16 Nov 1272. **Buried:** Westminster Abbey.

The nine-year-old Henry came to the throne at an unfortunate time, with a rebellion in full swing. Moreover, he had to be crowned with his mother's torque since his father had lost the crown. Things hardly improved when he grew old enough to rule, in 1232. By the Treaty of Paris (1259) he finally surrendered his claims to Normandy and Anjou, and did homage to Louis IX of France for Gascony. At home, opposition was provoked by his choice of advisers (too many foreign relatives) and excessive expenditure. The barons took the government into their own hands, forcing the king to accept the Provisions of Oxford (1258). When he reneged, they rebelled (1264), led by Simon de Montfort, Earl of Leicester. After a brief 'reign', Simon was defeated

Henry presiding at Parliament

by Henry's son, Edward, at Evesham (1265). With Henry growing senile, Edward became the effective ruler.

PARLIAMENT

The king's Great Council became known as 'parliament' during Henry's reign. Its growth resulted from the Crown's need for tax revenue. Representatives of the shires and the towns attended the parliament summoned during the Barons' War by Earl Simon in 1265.

A KINGDOM FOR A SON

A prime example of Henry's misjudgement was his scheme to make his second son, Edmund, King of Sicily (1254). Requiring vast expenditure – and a campaign to oust the existing king – it appeared highly impractical and provoked the barons into taking over the government in 1258.

♛ Though an inept ruler, Henry III was a cultured man and a patron of the arts, initiating the rebuilding of Edward the Confessor's Abbey of Westminster in glorious style.

♛ By his own wish, Henry was buried in the coffin of Edward the Confessor, its original incumbent having been removed to a grander one.

Henry III being crowned ▶

Edward I

Born: Westminster, 17 June 1239. **Parents:** Henry III and Eleanor of Provence. **Ascended the Throne:** 20 Nov 1272. **Coronation:** Westminster Abbey, 19 Aug 1274. **Authority:** King of England, Wales, Scotland and Ireland. **Married:** (1) Eleanor, daughter of Ferdinand III of Castile, (2) Margaret, daughter of Philip III of France (1299). **Children:** With (1), four sons, including the future Edward II, and 12 daughters; with (2), two sons and one daughter. **Died:** Burgh-by-Sands, Cumbria, 7 July 1307. **Buried:** Westminster Abbey.

Edward revived the dynasty after two ineffectual kings. King in all but name since 1265, Edward was returning from a crusade when his father died. With few French possessions except Gascony, he concentrated on asserting his sovereignty in Britain. Eight years' campaigning in Wales (1276-84) ended Welsh independence, reinforced by an unparalleled programme of castle-building to inhibit future revolts. In 1290 he asserted his overlordship of Scotland, and embarked on a series of domineering campaigns, earning the name 'Hammer of the Scots', confiscating the

Edward I with monks and bishops

Stone of Scone, and provoking a long, disastrous war.

- In 1273 Edward did homage to Philip III 'for all the lands which I ought to hold' in France, an ambiguous oath.
- At Caernarfon in 1284, according to legend, Edward proclaimed his only living new-born son, 'Prince of Wales', pointing out that he 'spoke no English'.
- In 1290 Edward expelled the Jews from England.

ROYAL JUSTICE

A forceful ruler, Edward established law and order in England, curbing the power of Church and barons and raising taxes through parliament. The so-called Model Parliament of 1295 was the most widely representative body yet summoned. The developing courts of the King's Bench, and Common Pleas furthered royal justice.

EDWARD AND ELEANOR

Edward was devoted to his wife Eleanor and when she died, near Grantham in 1290, he erected twelve memorial crosses marking the route of her cortège to London. Three originals (as opposed to replicas) still stand. His second marriage was also happy. Edward produced more legitimate children than any other monarch.

The siege of Berwick during Edward's Scottish campaign ▶

Edward II

Born: Caernarfon, 25 April 1284. **Parents:** Edward I and Eleanor of Castile. **Ascended the Throne:** 8 July 1307. **Coronation:** Westminster Abbey, 25 February 1308. **Authority:** King of England, other claims. **Married:** Isabella, daughter of Philip IV of France. **Children:** Two sons, including the future Edward III, and two daughters. **Died:** Berkeley Castle, 21 September 1327. **Buried:** Gloucester Cathedral.

Eccentricity was permissible in a successful monarch, but Edward's idiosyncrasies far outweighed his statecraft. His assumed lover, Piers Gaveston, a Gascon, irritated the English court, and Gaveston's murder by his enemies in 1312 was widely applauded. It turned Edward irrevocably against the ruling class. His next favourites, the Despensers (father and son), were more acceptable, but their intrigues against the Queen induced her withdrawal to her native France. She returned in 1326 with her lover Roger Mortimer and an army. The Despensers were captured and executed, Edward was imprisoned, forced to abdicate and later murdered.

SCOTTISH INDEPENDENCE

Edward belatedly continued his father's war against the Scots. Though not unathletic (reputedly a good swimmer), he was no warrior but was not present when his army was annihilated by the Scots, led by Robert Bruce, at Bannockburn in 1314. The battle assured Scottish independence.

A GRIM END

Edward was murdered in a peculiarly nasty manner. A horn was inserted into his rectum, and a red-hot spit thrust through it. His corpse was unmarked. Reports of this beastly act caused a belated surge of popular sympathy for Edward and hatred for Isabella – the 'She-Wolf of France' – who spent the rest of her life confined in Castle Rising (near King's Lynn).

♛ Edward's difficulties after 1314 were aggravated by poor harvests and epidemics among livestock.

♛ Naïve and tactless, it was said that Edward enjoyed humble company and had even helped dig ditches when he should have been at Mass.

♛ Edward's forced abdication set an ominous precedent.

Edward III

Born: Windsor Castle, 13 Nov 1312. **Parents:** Edward II and Isabella of France. **Ascended the Throne:** 25 Jan 1327. **Coronation:** Westminster Abbey, 29 Jan 1327. **Authority:** King of England, other claims. **Married:** Philippa, daughter of the Count of Hainault. **Children:** Eight sons, including Edward the Black Prince and John of Gaunt, and five daughters. **Died:** Sheen Palace, Surrey, 21 June 1377. **Buried:** Westminster Abbey.

Edward was a warrior-king in his grandfather's mould. In spite of heavy taxation to fund war, he developed into a genial, pragmatic and popular monarch, but his determination to re-establish the greatness of his dynasty led the Crown into the long pursuit of a futile goal – the throne of France. Edward's claim through his mother was not recognized in France. England became embroiled in the so-called Hundred Years' War – actually a series of wars from 1337 to 1453. Naval victory at Sluys (1340) gave England control of the Channel. The English were victorious at Crécy (1346) and Poitiers (1356), where they were led by Edward's eldest son, the Black Prince (1330-76).

♛ In 1340 Edward assumed the title, maintained by his successors until 1801, of King of France.
♛ The only permanent gain from Edward's French wars was Calais.
♛ In 1348-50 about one-third of the population of England died of bubonic plague in the Black Death.

LORDS AND COMMONS

Certain English institutions took recognizable form during Edward III's reign. Parliament was divided into two houses, and the procedure of impeachment was used against corrupt or incompetent ministers. Edward founded the Order of the Garter (1348), justices of the peace acquired more formal status, and English gradually replaced French as the 'official' language.

ALICE PERRERS

After the death of the amiable Queen Philippa (1369), Edward acquired a mistress, Alice Perrers, who shared his bed with her daughter. She is said to have infected him with gonorrhoea and, when he died after a stroke, to have stripped his body of jewels.

Richard II

Born: 6 Jan 1367. **Parents:** Edward the Black Prince and Joan, the 'Fair Maid of Kent'. **Ascended the Throne:** 22 June 1377. **Coronation:** Westminster Abbey, 16 July 1377. **Authority:** King of England. **Married:** (1) Anne of Bohemia, daughter of the Emperor Charles IV, (2) Isabella, seven-year-old daughter of Charles VI of France (1396). **Children:** None. **Died:** Pontefract Castle, 14 February 1400. **Buried:** King's Langley, removed to Westminster Abbey in 1413.

Richard, known as Richard of Bordeaux, succeeded his grandfather when aged ten. The chief power behind the throne was his uncle, John of Gaunt. The young king showed acumen and courage in confronting Wat Tyler, a leader of the Peasants' Revolt (1381), but lacked the capacity to control his uncles and other great lords. He antagonized many by arbitrary and unpredictable decisions and hefty poll taxes. While the king was in Ireland in 1399 John of Gaunt's son, Henry Bolingbroke, whom he had banished, invaded northern England and rallied enough support to force Richard's abdication in his favour (1399). Richard died a prisoner the following year, either through self-starvation or as a victim of murder.

That the king should be succeeded by his eldest son had become well established. Primogeniture reduced conflict over the succession, but sometimes placed a child on the throne. The unfortunate Richard was overshadowed by the formidable sons of Edward III, all too ambitious to be a father substitute.

CULTURE AND COOKERY

Richard, who is said to have invented the handkerchief, was passionately interested in dress, fine cooking and books. He presided over an exotically luxurious court, very different from his predecessors', at the centre of a remarkable flowering of English literature and art. The finest works of Geoffrey Chaucer, William Langland and John Gower were written during Richard's reign.

♛ Criticism of the king's advisers, meaning the royal government in general, reached a peak in the 'Merciless Parliament' (1388).

♛ Richard's court cookery book, called *The Forme of Cury,* contained recipes for dishes such as oysters in Greek wine.

♛ When Queen Anne died in Sheen Palace in 1394, Richard's grief and rage were such that he had the building levelled.

Richard taken to the Tower ▶

Henry IV

Born: Bolingbroke Castle, 4 April 1366. **Parents:** John of Gaunt and Blanche of Lancaster. **Ascended the Throne:** 30 Sept 1399. **Coronation:** Westminster Abbey, 13 Oct 1399. **Authority:** King of England. **Married:** (1) Mary de Bohun, daughter of the Earl of Hereford, (2) Joan, daughter of Charles II of Navarre. **Children:** With (1), five sons, including the future Henry V, and two daughters. **Died:** Westminster Abbey, 20 March 1413. **Buried:** Canterbury Cathedral.

Henry Bolingbroke succeeded his father (died 1399) as Duke of Lancaster and overthrew Richard II, who had seized the Lancastrian estates. As king, he was – predictably as a usurper – confronted by revolt and conspiracy himself. More energetic and more conciliatory than Richard, he overcame all opponents, defeating supporters of Richard II (1399), the powerful Percys of Northumberland (1408), and, most formidably, Owain Glyndŵr in Wales. The French took the opportunity to raid the south, and the Scots the north, but King James I of Scots was captured in 1406 and a truce with France agreed in 1407. Thereafter, Henry was relatively safe.

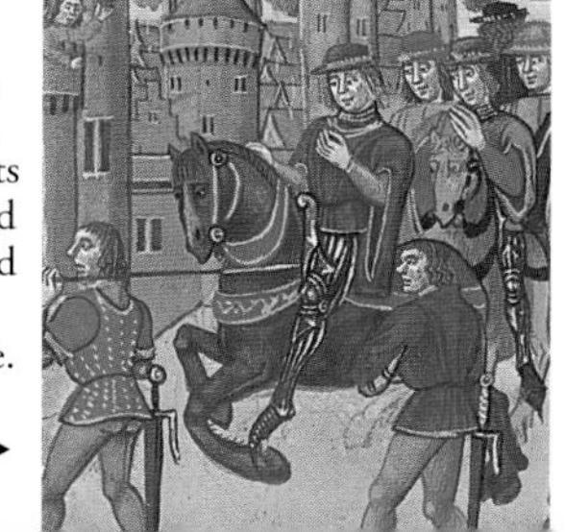

The young Henry Bolingbroke ▶

To the insecurity of the Crown was added the constant problem of insufficient income. Henry endured considerable criticism from parliament before it would grant the taxes he needed. His moderation and willingness to compromise ensured that, though he might not get all he wanted, he did not provoke dangerous opposition.

ROYAL LEPER?

Short and plain, presenting a marked contrast with the elegant Richard II, Henry suffered poor health in later years and was worn out at 47. He suffered from a disfiguring skin complaint which contemporaries thought was leprosy. Medical historians now believe it was a severe form of eczema.

♛ Henry IV was the first king of the House of Lancaster. There was no dynastic break, he and his predecessor both being grandsons of Edward III.

♛ It had been prophesied that Henry would die in Jerusalem. He presumed that meant on crusade. He actually died in the abbot's parlour at Westminster Abbey, a room known as the Jerusalem Chamber.

♛ Henry IV's greatest achievement was to leave his son a kingdom that was peaceful, loyal and united.

Coronation of Henry IV ▶

Henry V

Born: Monmouth, 9 August 1387. **Parents:** Henry IV and Mary de Bohun. **Ascended the Throne:** 20 March 1413. **Coronation:** Westminster Abbey, 9 April 1413. **Authority:** King of England, Duke of Normandy (from 1417), Regent of France (1420). **Married:** Catherine de Valois, daughter of Charles VI of France (1420). **Children:** One son, the future Henry VI. **Died:** Vincennes, 31 August 1422. **Buried:** Westminster Abbey.

Henry was the last, perhaps the ablest, of the medieval warrior-kings. He made England once more a continental power. Reviving the French wars, he won a sensational victory at Agincourt (1415). In 1417-19 he reconquered Normandy, advanced to Paris, and signed the Treaty of Troyes (1420), gaining Charles VI's daughter and recognition as regent and heir to the French throne. This was not popular in France, but Henry died, perhaps fortunately, before his grandiose plans of conquest could be fulfilled.

Henry and Catherine's marriage

MILITARY GENIUS?

Fearless and determined, Henry was also an able strategist, and the success of his French campaign was based on two years of careful planning. He took care to ensure control of the Channel and to strengthen his position by useful alliances, especially with Burgundy.

A POPULAR AUTOCRAT

Henry's single-minded pursuit of conquest now seems unattractive, and he was undoubtedly ruthless and authoritarian. Yet he was a hero. War was a popular activity, if you won. Henry encountered minimal opposition from England's turbulent nobility and raised extra revenue through parliament without serious difficulty.

- Henry's victory against heavy odds at Agincourt has been ascribed to the superiority of the longbow, allegedly a Welsh invention.
- Henry, personally devout, followed his father in persecuting the Lollards, Protestant dissidents.
- Henry died of dysentery, which probably caused more casualties in medieval armies than human foes.

Morning of Agincourt ▶

Henry VI

Born: Windsor Castle, 6 Dec 1421. **Parents:** Henry V and Catherine de Valois. **Ascended the Throne:** 1 Sept 1422. **Coronation:** Westminster Abbey, 6 November 1429; repeated St Paul's Cathedral, 13 Oct 1470. **Authority:** King of England. **Married:** Margaret, daughter of the Count of Anjou. **Children:** One son, Edward (died 1471). **Died:** Tower of London, 21 May 1471. **Buried:** Chertsey Abbey; removed to St George's Chapel, Windsor, 1484.

Henry presided over his first parliament from his mother's arms. Although he was crowned King of France in Paris (1431), the conquests of Henry V were soon lost. By 1453 only Calais remained. Henry was incapable of leading an army and, partly perhaps as a result of the French disasters, suffered his first attack of insanity in 1452. Richard, Duke of York reigned as Protector until Henry recovered in 1455, when the rivalry between Lancaster and York was developing into civil war. Henry was captured in 1460, deposed in 1461, briefly restored by Richard Neville, the Earl of Warwick, 'the Kingmaker', in 1470, before being finally returned to the Tower where he was stabbed to death by an unknown hand.

Henry's second coronation

THE MAID OF ORLÉANS

The French recovery was inspired by Joan of Arc and the coronation of Charles VII at Reims (1429). The English began to suffer defeats, their allies deserted them, and parliament had to provide ever-growing sums for what had become an unsuccessful war. Peace proved elusive. Even Henry's marriage to a formidable French princess in 1445 brought only a breathing space.

'A PREY UNTO THE HOUSE OF YORK'

Henry VI was pious and well-meaning. He might have made an excellent bishop, but kingship was beyond him. He lacked his father's qualities of leadership, and allowed power to fall into the hands of an inadequate clique. His attacks of insanity may have been inherited from his maternal grandfather.

- A rising in Kent led by Jack Cade (1450), partly provoked by dislike of the King's favourites, succeeded in taking London while Henry cowered in Kenilworth Castle.
- Henry's cast of mind was prudish. When shown his baby son, he expressed surprise, remarking that he must have been conceived by the Holy Ghost.

The Wars of the Roses

The dynastic conflict that began in the 1450s was later called the Wars of the Roses (the red rose of Lancaster versus the white rose of York). Beginning as a revolt against weak government, it became a challenge to the throne when Richard, Duke of York (1311-60) formed an alliance with the powerful Earl of Warwick. The Yorkists won the Battle of Towton (1461) and Richard's son, Edward IV, became king. Warwick, dissatisfied with his rewards, sided with the Lancastrians in 1469, and Henry VI was briefly restored (1470-1). The efforts of Richard, Duke of Gloucester, Edward IV's younger brother and guardian of the younger Edward V, to seize the crown himself alienated Lancastrians as well as Yorkists, and his overthrow by Henry Tudor (1485) marked the end of the Wars of the Roses.

In over 30 years, a state of open war existed for only about 15 months, and social disruption, except among the higher nobility, appears to have been minimal.

The leader of the Lancastrian cause, given the feebleness of Henry VI, was his tigerish queen, Margaret of Anjou (1429-82). After Towton she fled to France, returned with a French army, was defeated but returned again with more French soldiers, forcing Edward IV, in turn, to flee. Imprisoned in 1471, she was ransomed in 1475 and left England for good.

Edward IV in council

Edward IV with Elizabeth Woodville, Edward V and Richard Duke of Glucester

Edward IV

Born: Rouen, 28 April 1442. **Parents:** Richard, Duke of York, and Cecily, daughter of Ralph Neville, Earl of Westmorland. **Ascended the Throne:** 4 March 1461. **Coronation:** Westminster Abbey, 28 June 1461. **Authority:** King of England. **Married:** Elizabeth, daughter of Richard Woodville, Lord Rivers. **Children:** Three sons, including the future Edward V, and seven daughters; four illegitimate children. **Died:** Westminster, 9 April 1483. **Buried:** St George's Chapel, Windsor.

As soon as he was old enough, Edward fought in the Yorkist cause. After his father's death at Wakefield (1460) and the victory of Towton (1461), Henry VI was deposed and Edward declared king. Following the alliance of Margaret of Anjou and the Earl of Warwick, Edward was himself deposed (October 1470) and fled abroad but, with a small army of mercenaries, he regained the throne in April 1471. It took the death of Warwick at the Battle of Barnet, victory at Tewkesbury and the death of Henry VI and his only son to secure Edward's throne. Popular and pleasure-loving, he was nevertheless a formidable monarch, whose potential was extinguished by early death.

DREAMS OF GLORY

Well served by able ministers and in tune with England's increasingly prosperous merchants, Edward conciliated parliament by abjuring special taxes and built up commercially useful foreign connections. When his attempt to revive the claim to France failed, he accepted a handsome payment for withdrawing his forces.

'TOO GOOD TO BE HIS HARLOT'

• Edward, rare for a monarch, married for love and got away with it. The beautiful widow, Elizabeth Woodville (*c.*1437-92), who came from a family of Lancastrian gentry, refused to be his mistress so he married her secretly (1464), revealing the *fait accompli* only when he came under pressure to make a diplomatic match.

• Edward's brother George, Duke of Clarence, who joined Warwick and the rebels in 1469, was executed (or according to legend drowned in a butt of wine) in 1478.

• Edward's marriage and the lavish advancement of his wife's relations caused trouble and encouraged Warwick, 'the Kingmaker', to change sides.

Edward landing in Calais

Edward V

Born: Westminster, 4 November 1470.
Parents: Edward IV and Elizabeth Woodville.
Ascended the Throne: 9 April 1483.
Coronation: Not crowned. **Authority:** King of England.
Married: Unmarried. **Died:** Probably the Tower of London, *c.*September 1483. **Buried:** Tower of London; possible remains reburied in Westminster Abbey, 1678.

During his father's reign the young Prince of Wales was placed in the care of his mother's relations in Ludlow. Succeeding his father at the age of twelve, he was brought to London by his ruthless and ambitious uncle, Richard, Duke of Gloucester, his official guardian and Protector of the Realm. Leading members of the Woodvilles were arrested, and Edward, with his younger brother Richard, was lodged in the Tower. Prompted by Gloucester, parliament declared that Edward IV's marriage to Elizabeth Woodville had been invalid, and its issue therefore illegitimate. The young king was deposed on 25 June 1483. He and his brother were seen in the Tower in September, but never again.

THE DANGERS OF FACTION

The dynastic conflicts of the 15th century were largely the result of royal dependence on factions, which inevitably made enemies of those outside the magic circle. The Yorkist princes, including Edward IV, tended to behave like great lords rather than national leaders.

THE PRINCES IN THE TOWER

The fate of the the young princes poses the most famous conundrum in English history. There seems little doubt that they were murdered and – although interesting cases have been made against others – that their uncle Richard III was responsible.

♛ There was nothing particularly sinister in the choice of the Tower as the Princes' lodging. It was then a royal residence as well as a prison.

♛ In 1678 bones believed to be those of the murdered princes were discovered in the Tower. They were reinterred in Westminster Abbey on the order of Charles II.

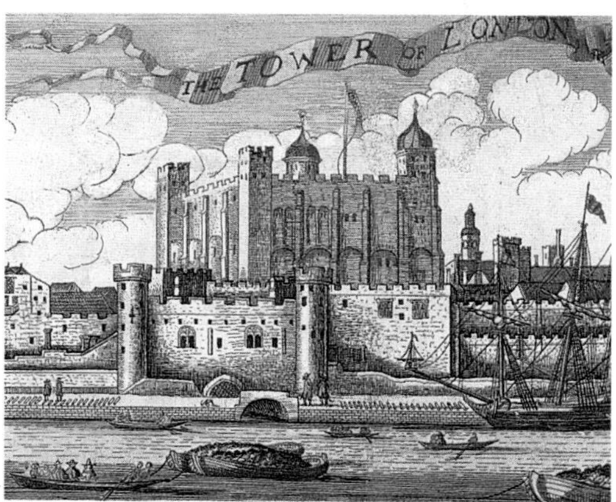

Tower of London

Richard III

Born: Fotheringhay Castle, 2 October 1452.
Parents: Richard, Duke of York, and Cecily Neville.
Ascended the Throne: 26 June 1483.
Coronation: Westminster Abbey, 6 July 1483.
Authority: King of England. **Married:** Anne Neville, daughter of the Earl of Warwick, 'the 'Kingmaker'.
Children: One son, Edward (died 1484); several illegitimate children. **Died:** Bosworth, Leicestershire, 22 August 1485.
Buried: Greyfriars Abbey, Leicester; later disinterred and desecrated.

A loyal and – as Duke of Gloucester and married to a Neville – powerful supporter of his brother, Edward IV, Richard served effectively in the Borders, recapturing Berwick from the Scots. On Edward's death, he ruthlessly suppressed the Woodvilles, forcing the widowed Elizabeth to take sanctuary at Westminster, and secured the Crown for himself. He antagonized many Yorkists, including the Earl of Buckingham (executed, with other opponents, without trial in 1483). Confronted by the Lancastrian invasion force of Henry Tudor at Bosworth, and deserted at the last minute by Lord Stanley with 7000 men, he was defeated and killed.

THE LAST YORKIST

In his brief reign Richard showed himself an able administrator, prepared to be conciliatory, more active than his predecessor but, like him, a patron of the arts.

THE WICKED UNCLE

Richard's reputation suffered through Tudor propaganda, much of it demonstrably false, and augmented by Shakespeare. He was a product of the degeneration in political behaviour since *c.*1450 and arguably no more ruthless than other Renaissance monarchs. Since the eighteenth century, he has had many defenders, and today his memory is rehabilitated by the Richard III Society.

- In spite of his nickname, 'Crouchback', Richard's disability seems to have been no more than a minor irregularity of the shoulder.
- Staring defeat in the face at Bosworth, Richard refused to flee, declaring, 'I will die King of England.'
- During the Reformation, Richard's bones were dug up and thrown into the River Soar.

Tudors & Stuarts

The Tudors

The claim to the throne made by Henry VII, the first monarch of the Tudor dynasty, was tenuous: Henry's mother was descended from John of Gaunt, a son of Edward III and ancestor of the House of Lancaster. Henry also claimed descent from Welsh princes. Compared with earlier aspirants, he was fortunate in the absence of other claimants. Richard III was dead and had no surviving children. Many Yorkist supporters,

Queen Elizabeth I

including Edward IV's widowed queen, Elizabeth Woodville, supported him, and the ruling class, after three decades of conflict, were war-weary. Henry portrayed himself as a figure of reconciliation, marrying a Yorkist princess, but, like other conquerors, he won and held the throne by his own efforts.

The House of Tudor, neatly encapsulating the 16th century by ruling from 1485 to 1603, was perhaps the most successful in English history. Its three most significant members, Henry VII, Henry VIII and Elizabeth I, who together reigned for almost 106 of those years, were all, whatever their personal qualities, highly capable rulers, though their reputation is less high among recent historians. They also reigned in a fortunate period for European monarchy (exemplified on the Continent by the Valois/Bourbon and Habsburg dynasties), when the power of the Crown was less threatened by 'overmighty subjects' and not yet seriously threatened by parliament. Among other achievements, the Tudors re-established the hereditary principle, which had seemed in danger of collapse during the 15th century.

Elizabethan London

Henry VII

Born: Pembroke, 28 Jan 1457. **Parents:** Edmund Tudor, Earl of Richmond, and Lady Margaret Beaufort. **Ascended the Throne:** 22 Aug1485. **Coronation:** Westminster Abbey, 30 Oct 1485. **Authority:** King of England. **Married:** Elizabeth of York, daughter of Edward IV. **Children:** Four sons, including the future Henry VIII, and four daughters. **Died:** Richmond, Surrey, 21 April 1509. **Buried:** Westminster Abbey.

Henry spent most of his youth in Wales and in exile in Brittany. After his victory at Bosworth he rapidly consolidated his position. Three 'pretenders', Lord Lovel, Lambert Simnel and Perkin Warbeck, appeared against him but none commanded sufficient support, and Henry restored the reputation of the monarch as one who rules, not merely reigns. His basic methods were to enforce the law, especially against magnates who exceeded their rights, and to exploit the Crown's powers of patronage. He selected his closest advisers for their loyalty and ability, especially in raising Crown revenue. As he also avoided foreign wars, Henry never needed to appeal to parliament for funds and left a full treasury.

GROWING KINGDOM

Population was expanding, especially in London, indicating growing prosperity. Like his predecessors, Henry VII encouraged both trade, making advantageous commercial treaties, and the cloth industry. Among the enterprises he sponsored was the voyage from Bristol of John Cabot in 1497 which led to the discovery of the North America mainland.

RESPECTED THOUGH NOT LOVED

Henry is remembered for the magnificent Perpendicular interiors of his chapel in Westminster Abbey and King's College, Cambridge. But, if not the miser critics called him, he was tough and ruthless in pursuit of his interests, and some of his expedients for raising money from his subjects were of doubtful legality. As a result, he was not popular with his subjects, some of whom greeted his death with celebration.

- With confident irony, Henry put the disgraced pretender Lambert Simnel to work in the royal kitchens.
- Henry's mother, Lady Margaret Beaufort, was a notable patron of religious and educational foundations. Her name is still gratefully remembered by the universities of Oxford and Cambridge.
- Great things were expected from Henry's eldest son, Arthur, Prince of Wales, but he died in 1502.

Henry VIII

Born: Greenwich, 28 June 1491. **Parents:** Henry VII and Elizabeth of York. **Ascended the Throne:** 21 April 1509. **Coronation:** Westminster Abbey, 24 June 1509. **Authority:** King of England and Ireland. **Married:** Six wives (see page 170-1). **Children:** Two daughters and one son (see page 170-1); three or four illegitimate children. **Died:** Whitehall, 28 Jan 1547. **Buried:** St George's Chapel, Windsor.

Handsome, intelligent, athletic, Henry appeared the perfect Renaissance prince. He added to his popularity by executing his father's hated tax collectors, Richard Empson and Edmund Dudley, but until 1529 he left mundane administration largely in the hands of Thomas Wolsey, the ambitious and able son of a Suffolk grazier. Abandoning his father's peaceful foreign policy, Henry wasted resources on flamboyant but unsuccessful expeditions against France and wars with Scotland and Spain. He recouped vast amounts by dissolving the monasteries and confiscating their property (1536-9), but squandered most of that in military expenditure in the 1540s.

Henry installed himself as head of the Church

REFORMATION

As a result of difficulties in obtaining permission to

divorce from the Pope, Henry exploited widespread anticlericalism to end the authority of the Pope and place himself at the head of the English Church. This was accomplished in 1532-6 through a series of momentous Acts of Parliament, with the guidance of Thomas Cromwell, Wolsey's successor. One result of this revolution was to raise the constitutional status of parliament.

FROM PRINCE TO OGRE

Henry was a second son, distrusted by his father. An inferiority complex may have been responsible for his intense egoism and extreme touchiness, which, combined with declining health and increasing megalomania, turned him from the 'Bluff King Hal' of his early years into the frightening despot of the 1540s.

Ironically, Henry earned the title *Fidei Defensor*, 'Defender of the Faith', from the Pope for his pamphlet attacking Luther. The title is still held and appears, abbreviated, on modern British coins.

Greed made Henry grossly fat. At Boulogne in 1544 he was carried about in a chair and hauled upstairs by machinery.

The Wives of Henry VIII

Henry VIII is the best known of English monarchs, largely because he had six wives. This unrivalled record was partly due to his character, but also to political demands and chance circumstance, such as the need to produce healthy boys. For all Henry's efforts, only three legitimate children survived him, only one of them a boy, and a sickly youth at that.

(1) **Catherine of Aragon** (1485-1536), daughter of Ferdinand II of Aragon and Isabella of Castile; widow of Prince Arthur, Henry's elder brother. She married Henry on 11 June 1509. Although the queen conceived eight times, no child survived infancy except **Mary.** By *c.*1525 it was obvious that Catherine would not produce a male heir. The Pope's refusal to grant an annulment set in train the events leading to the break with Rome.

(2) **Anne Boleyn** (*c.*1507-36), daughter of Sir Thomas Boleyn and Lady Elizabeth Howard. She was pregnant when Henry married her on 25 January 1533. Besides **Elizabeth,** she had two still-born children and one miscarriage. Failure to produce a boy and accusations of adultery, probably just foolish flirtation, led to her execution for treason on 19 May 1536.

◀ *Catherine of Aragon*

(3) **Jane Seymour** (*c.*1505-37), daughter of Sir John Seymour and Margaret Wentworth. She married Henry on 30 May 1536 and died soon after giving birth to a son, **Edward.**

(4) **Anne of Cleves** (1515-57), daughter of the Duke of Cleves. She married Henry on 6 January 1540. It was an arranged, diplomatic match and Anne proved to be much plainer than Holbein's pre-nuptual portrait of her had suggested. The marriage was annulled on 9 July 1540.

(5) **Catherine Howard** (*c.*1520-42), daughter of Lord Edmund Howard. She married Henry on 28 July 1540. As a high-spirited girl tied to a bloated and repellent husband, she may have been guilty, as accused, of infidelity, for which she was executed on 13 February 1542.

(6) **Catherine Parr** (*c.*1512-48), daughter of Sir Thomas Parr. A sensible lady already twice widowed, she married Henry on 12 July 1543. Her function was that of nurse and stepmother, roles she fulfilled admirably.

The fate of Henry's wives is commemorated in a mnemonic:

Divorced, beheaded, died,
Divorced, beheaded, survived.

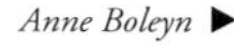

Anne Boleyn ▶

Edward VI

Born: Hampton Court, 12 Oct 1537. **Parents:** Henry VIII and Jane Seymour. **Ascended the Throne:** 28 Jan 1547. **Coronation:** Westminster Abbey, 19 Feb 1547. **Authority:** King of England and Ireland. **Married:** Unmarried. **Died:** Greenwich, 6 July 1553. **Buried:** Westminster Abbey.

As Edward was only nine when his father died, his uncle, Edward Seymour, later Duke of Somerset, became Lord Protector until, having failed disastrously to carry through Henry VIII's aggressive schemes in Scotland, he was displaced in 1550 (executed 1552) by John Dudley, Earl of Warwick, who appointed himself Duke of Northumberland. Always frail, Edward, whose name is commemorated in numerous grammar schools founded during his reign, died at 15, probably of tuberculosis.

REFORMATION CONTINUED

During Edward's reign, the Henrician Reformation was carried through to its logical conclusion when the English Church adopted Protestantism with the introduction of the

First Book of Common Prayer (1549), compiled by Thomas Cramner, Archbishop of Canterbury.

PRECOCIOUS YOUTH

Edward was intensively educated by leading scholars, who inclined him towards the reformed religion and turned him into a learned young pedant. He enjoyed a short family life with his sisters under the amiable tutelage of Catherine Parr.

THE 'NINE DAYS' QUEEN

In pursuit of his family ambitions, the Duke of Northumberland persuaded Edward to will the Crown to **Lady Jane Grey** (1537-54). A Protestant, she was the granddaughter of Henry VIII's sister, Mary, and was married to Northumberland's son, Lord Guildford Dudley. She was proclaimed queen on Edward's death. Northumberland's scheme fell flat when neither Lords nor Commons would accept her and declared for the rightful heir, Mary. Jane and her husband were executed on 12 February 1554. Northumberland, who unlike Jane had recanted in an effort to save his head, preceded them to the block in 1553.

Mary I

Born: Greenwich Palace, 8 February 1516.
Parents: Henry VIII and Catherine of Aragon.
Ascended the Throne: 19 July 1553.
Coronation: Westminster Abbey, 1 October 1553.
Authority: Queen of England and Ireland.
Married: Philip II of Spain. **Children:** None.
Died: Whitehall, 17 November 1558.
Buried: Westminster Abbey.

Half-Spanish and a devout Roman Catholic, Mary married the future King of Spain, Philip II, as his second wife, in 1554. The marriage was unpopular and provoked Wyatt's Rebellion. Mary's great aim was to restore the authority of the Pope and return the country to Roman Catholicism. She earned her nickname 'Bloody Mary' for persecution of Protestants, about 280 of whom were burned. Some of these, such as Archbishop Thomas Cranmer, were political victims; others were militants of a kind who would have been burned under Henry VIII. Mary's failure to produce an heir assured a Protestant future.

CALAIS LOST

In 1557 Philip, now King of Spain, made a short visit to England, enlisting the country in his war against France. The result was the loss of Calais (1558), the last remaining

English possession in France. Mary said that when she died the words 'Philip' and 'Calais' would be found engraved upon her heart.

AN UNHAPPY UNION

Mary, who had once been courted by her husband's father, Charles V, was devoted to Philip though he found her repellent. She insisted that he should have the title of king and, although he spent more time in Spain, both their names appeared on Acts of Parliament and both their faces on the coinage. So desperate was she for a child that in 1555 she convinced herself she was pregnant.

♛ Execution by burning, a particularly cruel method, was reintroduced in February 1555.

Elizabeth I

Born: Greenwich, 7 September 1533.
Parents: Henry VIII and Anne Boleyn.
Ascended the Throne: 17 November 1558.
Coronation: Westminster Abbey, 15 January 1559.
Authority: Queen of England and Ireland.
Married: Unmarried. **Died:** Richmond Palace, Surrey, 24 March 1603. **Buried:** Westminster Abbey.

Elizabeth, England's most popular ruler, had a difficult childhood, having been declared illegitimate after the fall of Anne Boleyn. Under Mary she was a prisoner, held briefly in the Tower, as a likely focus of Protestant plots. She proved to be a ruler of quality: courageous, shrewd and possessing a potent way with words, although she was politically indecisive. Historically she benefited from the extraordinary cult of 'Gloriana' created around her by courtiers, and from the exceptional quality of Elizabethan mariners, poets (including Shakespeare) and other Renaissance heroes who ornamented her long and prosperous reign. Her aim was stability and concord, but administration was neglected. Crown revenue declined, corruption crept into government, and a rift began to open between Crown and parliament.

THE THIRTY-NINE ARTICLES

Elizabeth's major achievement was the settlement of the religious question, with the creation of the Church of England, based on the Thirty-Nine Articles (1563).

However, it automatically turned English Catholics into traitors, and displeased those radical Protestants who came to be known as Puritans.

THE VIRGIN QUEEN

Although Elizabeth had many suitors, and was romantically involved with Sir Robert Dudley, Earl of Leicester, she never married. She was able to exploit her availability to some advantage, and it augmented her popularity, always carefully courted, with the common people. Yet spinsterhood may have been due to personal preference as well as statecraft.

♛ Philip II of Spain, England's former king, now enemy, launched an invasion fleet, the famous Spanish Armada in 1588, but it was broken up by the navy and storms.

♛ Elizabeth's portraits show not so much a woman or even a queen as a gorgeous icon.

♛ Elizabeth kept the note Leicester had written to her from Rycote a few days before his death in a little casket by her bed, where it was found after her own death, fifteen years later. Across it she had written 'his last letter'.

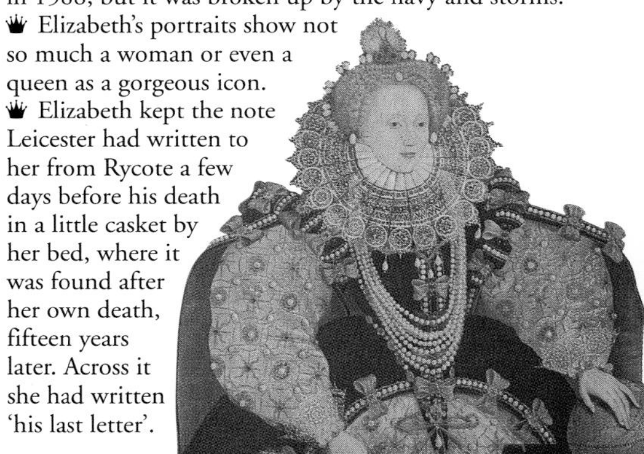

The Stuarts

The Stewart, or Stuart, dynasty, long established in Scotland, was not very successful in England, especially by comparison with the Tudors. In spite of inheriting a Crown that seemed more secure than any in Europe, two of the four Stuart kings lost it, and one of them, Charles I, lost his head as well. For a time (the first – and last – time since the Anglo-Saxons), Britain had no monarchy at all.

The Stuart monarchs all had defects of character. In some cases those defects were not only glaringly apparent but politically dangerous. All the same, the unfortunate record of the dynasty cannot be ascribed to such simple causes alone.

Although the Stuart period appears dominated by the civil conflicts of the middle decades of the 17th century, great though less obvious changes were taking place in society – changes, some historians argue, of greater significance than the sensational events of the civil wars. The monarchy as inherited by the Stuarts commanded enormous powers, but it was ill-equipped to understand or regulate these changes, and the resources at its disposal were insufficient to carry out the tasks it was expected to perform. Thanks to taxes on trade, revenue was just sufficient to govern the country in peacetime, but not enough in time of war. Additional revenue could only be raised through parliament, but parliament was increasingly critical of royal government, and inclined to attach unacceptable conditions to its grants. Gradually, disagreement spiralled into open war.

◀ *James I*

James I

For James's early life and reign in Scotland, see James VI, King of Scots, page 52.

Ascended the Throne: 24 March 1603. **Coronation:** Westminster Abbey, 25 July 1603. **Authority:** King of Great Britain and Ireland. **Married:** Anne, daughter of Frederick II of Denmark and Norway (1589). **Children:** Three sons, including the future Charles I, and five daughters. **Died:** Theobalds Park, Hertfordshire, 27 March 1625. **Buried:** Westminster Abbey.

Scholarly but unkingly, James irritated some courtiers by his informal habits and homosexual inclinations. He soon came into conflict with Puritans in the Church and narrow-minded squires in parliament, and his vision of himself as a symbol of unity between his three kingdoms and between the Scottish and English Churches was unrealized. The outbreak of the Thirty Years War (1618) and war with Spain (1624) spoiled his efforts as a peacemaker, and the planting of Protestant Scots and English in Ulster stored up future troubles. But the kingdom was peaceful and secure, religious and other conflicts diminished, and the fundamental authority of the Crown was unchallenged.

TO VIRGINIA

In James's reign the first permanent English colonies were established in Virginia (named for the 'Virgin Queen'). Jamestown was sponsored by London merchants and founded in 1607. The first colony in

New England was established by the Pilgrim Fathers, a group of idealistic religious dissidents who sailed on the *Mayflower* in 1620.

ROYAL FAVOURITES

After the death of the capable royal servant, Robert Cecil, Lord Salisbury, in 1612, James tended to rely on favourites. They included the dubious Robert Carr, Earl of Somerset, who became involved in a murder case, and the handsome but pretentious George Villiers, Duke of Buckingham (assassinated 1628).

♛ An auspicious result of the Hampton Court conference on religious affairs (1604) was the commissioning of an authorized version of the Bible – the King James Bible.

♛ James's queen, Anne of Denmark, led a largely separate life, enjoying royal 'progresses' around the country.

♛ James's support for the episcopalian system of Church government was enshrined in his phrase, 'No bishop, no king'.

Anne of Denmark ▶

Gunpowder Plot

English Roman Catholics after the Anglican settlement of 1559 were faced with irreconcilable, dual loyalties. The Pope demanded the deposition of the heretic Elizabeth in favour of the Catholic Mary, Queen of Scots. That was the aim of the Northern Rebellion (1569). Some Catholics were persecuted and priests, regarded as secret agents, were executed. Other, Spanish-inspired, Catholic plots, resulted in the execution of Mary (1587).

The last 'Popish' plot took place in the reign of James I. It involved a group of Catholic gentlemen, mostly from the West Country. They were betrayed, and Guy Fawkes was arrested in the cellars of the House of Lords as he prepared to blow up king and parliament with 20 barrels of gunpowder on the evening of 4 November 1605.

Guy Fawkes was a Yorkshireman, a soldier of fortune who had been fighting with the Spanish army in the Netherlands. He was tortured to make him reveal the names of his co-conspirators. Their leader, Robert Catesby, was killed resisting arrest. The others, including Fawkes, were tried by special commission and executed on 1 February 1606.

Parliament named 5 November, the day the explosion was planned, as a day of public thanksgiving. It has since been celebrated with fireworks and the burning of a 'guy'.

Guy Fawkes before James ▶

Charles I

Born: Dunfermline, 19 Nov1600. **Parents:** James VI and I and Anne of Denmark. **Ascended the Throne:** 27 Mar 1625. **Coronation:** Westminster Abbey, 2 Feb 1626. **Authority:** King of Great Britain and Ireland. **Married:** Henrietta Maria, daughter of Henri IV of France. **Children:** Four sons, including the future Charles II and James II, and five daughters. **Died:** Whitehall, 30 Jan 1649. **Buried:** St George's Chapel, Windsor.

Charles was a frail and backward child who became heir apparent on the death of his promising elder brother, Henry, in 1612. Encountering opposition in parliament, he ruled without it during the 'Eleven Years Tyranny' (1629-40), levying taxes unsanctioned by parliament. A revolt by the Scots, provoked by measures to align the Presbyterian Kirk with the Church of England, forced him to call parliament to raise money. Exasperated by his illegal imposts, the Commons demanded redress of grievances

first. Charles's attempt to arrest five leading members resulted in his flight from London, which he never succeeded in regaining. This put him at a hopeless disadvantage throughout the conflict with parliament; he was executed for treason seven years later.

DIVINE RIGHT

In the contest between Crown and parliament, both sides had genuine grievances. A poor communicator, Charles was so confident of his 'divine right' to rule that he was not prepared to persuade, conciliate or compromise, and regarded all opposition as, by definition, treasonous. The Commons, understanding little of the problems of government, were equally unreasonable.

'ROYAL MARTYR'

The execution of the king shocked Europe, as well as most of his subjects. His dignity and courage restored royal prestige and inflicted a propaganda defeat on his opponents. Though he had been dishonest and unreliable, sacrificing loyal ministers to preserve his own position, he is still remembered as the royal martyr.

♛ Charles's patronage of the arts and appreciation of painting accounts for the large number of portraits of him.
♛ Charles, married to a Roman Catholic and sympathetic to the Arminian, high-church movement represented by Archbishop William Laud, was suspected of Roman Catholic sympathies.
♛ Charles wore two shirts to his execution because it was a cold day and he was afraid if he shivered people would think he was afraid.

The English Civil Wars

Following a Scottish invasion (1638), an Irish rebellion in Ulster (1641) and the break between king and parliament, war began in 1642. Parliament's control of London gave it a significant advantage. The royalists, with headquarters at Oxford, held most of the north and west. Early fighting favoured the royalists ('Cavaliers'), but Scottish aid and the new troops of **Oliver Cromwell** (1599-1658) secured the victory of Marston Moor (1644) for parliament ('Roundheads'). The Battle of Naseby (1645) was decisive. The royal forces disintegrated and Charles surrendered to the Scots, who handed him over to the army of Cromwell and Sir Thomas Fairfax (1647). His intrigues with all parties made agreement impossible and provoked a

Oliver Cromwell

second civil war, with the Scots now on Charles's side. Cromwell's New Model Army defeated them at Preston (1648). The trial and execution of the king, a 'cruel necessity' according to Cromwell, followed.

In 1649-51 Cromwell crushed support for Charles II in Scotland and Ireland. At Drogheda and Wexford his troops took brutal revenge for massacres of Ulster Protestants by Irish Catholics in 1641. The army was now the real power in the land.

Parliamentary government during the Interregnum (1649-60) proved no less arbitrary than the king's. Various expedients were tried before Cromwell, the outstanding statesman of his time, was made Lord Protector (1653), with the authority of a king. His son Richard succeeded him (1658) but lacked his authority. Charles II was invited to reassume the Crown in 1660.

Battle of Marston Moor

Charles II

Born: St James's Palace, 29 May 1630. **Parents:** Charles I and Henrietta Maria. **Ascended the Throne:** 30 Jan 1649 (by right); restored 29 May 1660. **Coronation:** As King of Scots, Scone, 1 Jan 1651; as King of England, Westminster Abbey, 23 April 1661. **Authority:** King of Great Britain and Ireland. **Married:** Catherine of Braganza, daughter of the King of Portugal. **Children:** Three children still-born; about 17 illegitimate children, including the Duke of Monmouth (1649-85). **Died:** Whitehall, 6 Feb 1685. **Buried:** Westminster Abbey.

After Cromwell's victory at Worcester (1651) Charles fled first to France and later to the Netherlands. He was recalled to the throne by parliament in 1660 at the instigation of General Monck, with powers scarcely less than those of his father and a determination not to 'go on his travels again'. He presided over an extravagant court, and supported revival of the theatre (banned during the Commonwealth) and the founding of the Royal Society (1660). His reign was marked by disasters such as the Great Plague (1665) and the Fire of London (1666), in which Charles helped fight the flames. In renewed conflict with the Dutch, part of the British fleet was destroyed when the Dutch sailed unchallenged up the Medway (1672).

MONEY PROBLEMS

Royal revenue continued to be insufficient, largely due to profligate spending. Charles circumvented the problem by deceiving his ministers and accepting secret subsidies from France, cynically promising to reintroduce Roman Catholicism (though he did die a Roman Catholic himself).

AN EXTENDED FAMILY

Charles had numerous mistresses, including the orange-seller Nell Gwynne, and bastards, several of them the ancestors of current dukes. One son, the Duke of Monmouth, became a rival to his Roman Catholic uncle, the future James II, and he attempted to seize the throne after Charles's death.

- The English coronation regalia had been broken up during the Interregnum. A new set had to be made in 1660.
- Charles and his brother James narrowly escaped assassination on their way from Newmarket races in the Rye House Plot (1683).
- The wittiest of British monarchs, Charles on his deathbed apologized to hovering courtiers for taking so long to die.

James II

Born: St James's Palace, 14 October 1633. **Parents:** Charles I and Henrietta Maria. **Ascended the Throne:** 6 Feb 1685. **Coronation:** Westminster Abbey, 23 April 1685. **Authority:** King of Great Britain and Ireland. **Married:** (1) Anne Hyde, daughter of Edward Hyde, Earl of Clarendon, (2) Mary, daughter of the Duke of Modena. **Children:** With (1), four sons and four daughters, including the future sovereigns Mary II and Anne; with (2), two sons, including James Edward the 'Old Pretender', five daughters and five still-born babies. **Deposed:** 23 Dec 1688. **Died:** Château de Saint Germain-en-Laye, near Paris, 6 Sept 1701. **Buried:** Saint Germain-en-Laye; later, the church of the English Benedictines in Paris; possibly returned to Saint Germain.

James had a good record as a soldier, naval commander and public servant. As an admitted Roman Catholic he was compelled to relinquish his offices by the Test Act (1673), but was gradually rehabilitated by Charles II, who resisted all attempts to promote a Protestant heir in James's place. The Duke of Monmouth, his nephew and Protestant rival to the throne, was defeated, but James's efforts to restore Roman Catholicism in England by packing parliament and dismissing opponents (including 75 per cent of JPs) aroused widespread antagonism, leading to his downfall.

The birth of a male heir (James Edward), implying a Roman Catholic succession, clinched James's fate. The rumour that the baby had been smuggled into the

Queen's bed in a warming pan was wishful thinking. The birth prompted an invitation to William of Orange, married to James's Protestant daughter Mary, to take the Crown. James fled to France.

A STIFF-NECKED MONARCH

James did not plan to make himself a continental-style despot, nor to destroy the Church of England. He merely to sought to restore equality for Catholics. But his mulish temperament, narrow vision and willingness to ride roughshod over opponents made people think the worst.

- After their defeat at the Battle of Sedgemoor (1685), supporters of Monmouth's Rebellion in the West Country (1685) were brutally dealt with by Judge Jeffreys in the 'Bloody Assizes'.
- A trade boom that boosted royal revenue reduced James's dependence on parliament.
- Fleeing London in December 1688, James dropped the Great Seal of England in the River Thames.

Mary II

Born: St James's Palace, 30 April 1662. **Parents:** James II and Anne Hyde. **Ascended the Throne:** 13 February 1689. **Coronation:** Westminster Abbey, 11 April 1689. **Authority:** Queen of England, Scotland and Ireland. **Married:** William of Orange. **Children:** None born alive. **Died:** Kensington Palace, 28 December 1694. **Buried:** Westminster Abbey.

Mary married her Dutch cousin at St James's Palace in 1677. In spite of physical disparity (William was small, Mary large), lack of children, and William's preference for male company, it was a very successful match. Mary supported her husband's insistence that he should be king rather than merely consort. She was entirely subservient to him in affairs of state, but deputized adequately when he was absent abroad. Unlike her sister Anne, Mary showed little sympathy for her father, whose throne she had usurped, partly because of their religious differences. Mary inclined to the Calvinism of Holland, and the court observed a stricter code of morals than had been customary since 1660.

SMARTENING UP THE PALACE

William and Mary did not like living in Whitehall, which was draughty and bad for the King's asthma. They preferred Hampton Court and, especially, Kensington Palace, both greatly expanded by Christopher Wren.

ROYAL SISTERS

William was responsible for reconciling Mary with her younger sister, Anne, with whom she was on bad terms for several years. He also pleased Anne when he appointed John Churchill, later Duke of Marlborough, and the husband of Anne's close friend, commander-in-chief, in spite of his involvement in intrigues against the throne. When Mary died, William reigned alone with Anne's approval.

♛ When Mary arrived in England to be proclaimed queen, it was reported by Sarah Churchill, who disliked both William and Mary, that she looked 'into every closet and conveniency, and turning up the quilts upon the bed, as people do when they come to an inn'. In reality, she had been warned that she must be cheerful and confident when she arrived in England, and in her enthusiasm, rather overdid it.

♛ Mary died of smallpox. William was with her when she died and was so devastated by grief that it was feared he might die too or lose his sanity. Though urged to marry again after Mary's death, William never did.

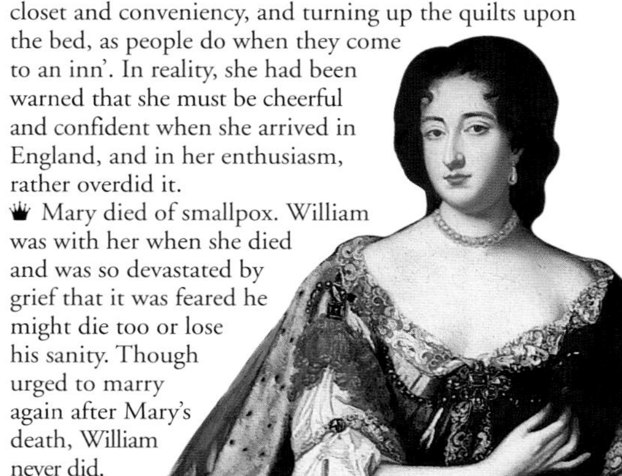

William III

Born: The Hague, 4 Nov 1650. **Parents:** Prince William II of Orange, Stadtholder of the Netherlands, and Mary Henrietta, daughter of Charles I. **Ascended the Throne:** 13 Feb 1689. **Coronation:** Westminster Abbey, 11 April 1689. **Authority:** King of England, Scotland and Ireland, Stadtholder of the Netherlands. **Married:** Mary, daughter of James II. **Children:** None born alive. **Died:** Kensington Palace, 8 March 1702. **Buried:** Westminster Abbey.

William was already Stadtholder, virtually a hereditary monarch, of the Netherlands. The throne was offered to him and Mary jointly, a unique arrangement, as Mary was the actual heir but William insisted on being king. He reigned alone after Mary's death. There was practically no resistance to William in England when he landed at Brixham, but the deposed James II invaded Ireland in 1689, provoking a bloody campaign that terminated in William's victory at the Battle of the Boyne in 1690. Supporters of James in Scotland were defeated at Killiecrankie and Dunkeld. William was a Dutch patriot with wide political horizons: his first commitment was to a European alliance against French aggrandizement. France became, and long remained, Britain's enemy too.

THE 'GLORIOUS REVOLUTION'

The revolution of 1688-9 was regarded as glorious because it was achieved without violence (in England anyway). The Bill of Rights, the constitutional settlement of 1689, represented the victory of parliamentary monarchy.

A PROTESTANT CHAMPION

William's lifelong resistance to the mighty Louis XIV of France made him the Protestant champion of Europe. Though physically small and not strong, he proved himself a good soldier as well as a diplomat. He was never, however, very popular in England.

William died a few days after a fall when his horse stumbled over a molehill. Jacobites used to toast the 'little gentleman in black velvet' responsible for the accident.

Archbishop Sancroft declined to crown William III, having crowned James II four years earlier, so the Bishop of London performed the ceremony.

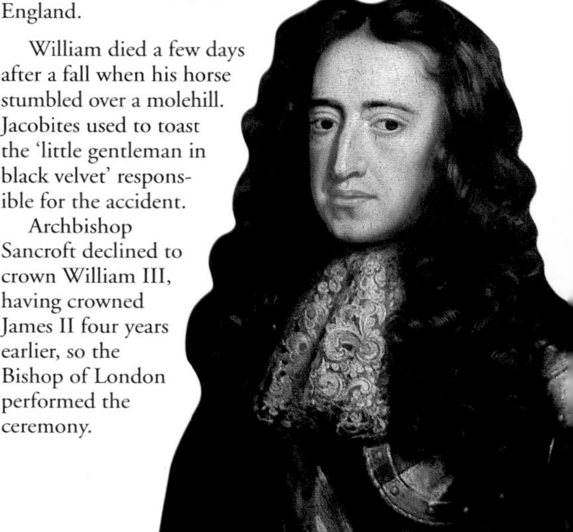

Anne

Born: St James's Palace, 6 Feb 1665. **Parents:** James II and Anne Hyde. **Ascended the Throne:** 8 March 1702. **Coronation:** Westminster Abbey, 23 April 1702. **Authority:** Queen of Great Britain and Ireland. **Married:** George, son of Frederick III of Denmark. **Children:** 18, including those still-born and miscarried. **Died:** Kensington Palace, 1 Aug 1714. **Buried:** Westminster Abbey.

Anne was an amiable and dutiful woman who presided over government without getting too involved in the intricacies of administration or the arguments of *Whigs* and *Tories,* though she favoured the latter and was devoted to the Anglican Church. She was not particularly intelligent, and her husband was notoriously dim, though amiable. Their sorrow was that, in spite of 18 pregnancies, none of Anne's children survived infancy except for one boy, who reached 11. Anne was given to intense female friendships, first with the Duchess of Marlborough, whose husband's victories over the French lent the Crown reflected glory, and later with Abigail Masham.

Prince George of Denmark ▶

THE UNITED KINGDOM

The Act of Union between England and Scotland (1707) made Anne the first sovereign of the United Kingdom. Though the Scots lost their parliament, they retained their own religious and legal systems. Union also reduced the danger of the Scots opting for a Jacobite succession.

A SQUARE COFFIN

Anne's health was poor. She suffered from gout and was unable to walk up the aisle at her coronation. Her ill luck as a mother may have caused her to over-eat – and drink. Like her sister Mary, she was large, and grew larger. Her coffin was said to be almost cubic in shape.

- Like her ancestor, Henry VIII, Anne was sometimes moved about with the aid of chairs and pulleys.
- The Duke of Marlborough was dismissed in 1711 by a Tory ministry resentful of the cost of the British involvement in the War of the Spanish Succession (1702-13).
- A Jacobite 'pretender', backed by France, made a sortie to Scotland in 1708, but was prevented from landing.

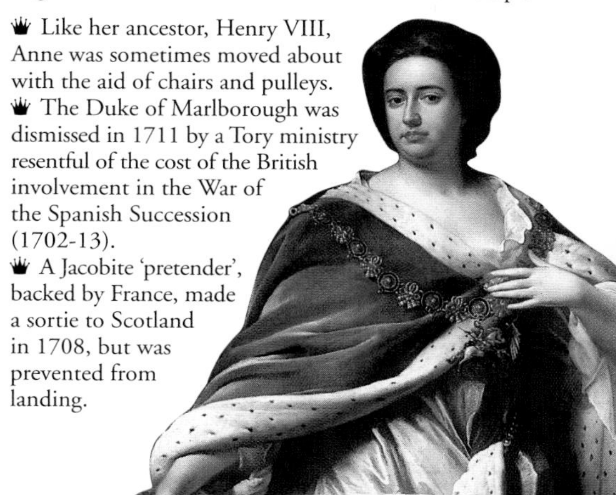

Hanover & to Windsor

The Hanoverians

When it became clear that Queen Anne, like her sister, would not provide an heir, the Act of Settlement was passed in 1701 to prevent a Jacobite restoration. It settled the succession on the elderly Electress Sophia of Hanover, who was the grand-daughter of James I through her mother, Elizabeth, wife of Frederick V, King of Bohemia (the 'Winter King'). She was, of course, a Protestant. As Sophia died two months before Anne, the Crown was inherited by her son George, Elector of Hanover since the death of his father in 1698.

Lack of familiarity with British affairs on the part of the first two Hanoverian kings assisted the gradual transfer of executive power from the monarch to the prime minister and cabinet.

Despite continuing threats from Jacobite claimants, who refused, unlike Henri IV of France, to change their religious allegiance in order to strengthen their dynastically superior claim, the Hanoverian line continued uninterrupted through five monarchs. In 1837 the Crown descended to Queen Victoria, six generations removed from George I, while Hanover, where female succession was forbidden, passed to a male descendant of George III. Victoria married Prince Albert of Saxe-Coburg-Gotha, whose family name was **Wettin,** in 1840. In 1917, when Britain was at war with Germany, George V adopted the name **Windsor** for his dynasty. Elizabeth II, though married to a Mountbatten (itself an Anglicization of the German 'Battenberg') perpetuated the name Windsor by proclamation but revised her decision in 1960 so that the third generation of her male descendants should bear the name **Mountbatten-Windsor.**

George I

Born: Osnabrück, Hanover, 28 May 1660. **Parents:** Ernest Augustus, Elector of Hanover, and Sophia, daughter of Elizabeth of Bohemia. **Ascended the Throne:** 1 Aug 1714. **Coronation:** Westminster Abbey, 20 Oct 1714. **Authority:** King of Great Britain and Ireland, Elector of Hanover. **Married:** Sophia Dorothea of Celle. **Children:** One son, the future George II, and one daughter; three illegitimate children. **Died:** Osnabrück, 11 June 1727. **Buried:** Leineschlosskirche, Hanover; reinterred Herrenhausen, Hanover, 1957.

George ascended the English throne with only a slight knowledge of English. He communicated with his ministers in French. He left his wife, from whom he was divorced, behind but brought to England two ladies promptly nicknamed, for obvious reasons, 'the Maypole' and 'the Elephant'. The first was his mistress, perhaps his wife (a secret marriage was rumoured), whom he made Duchess of Kendal. The other, though also assumed to be a mistress, was his half-sister, created Countess of Darlington. George preferred Hanover to his new kingdom and returned frequently. He never established much empathy with his British subjects.

BRITAIN'S FIRST PRIME MINISTER

George was inevitably dependent on his ministers, especially Robert Walpole, the Whig magnate and political leader often called the first prime minister – though the term was originally used mockingly. Walpole gained power through his skilful handling of the disastrous stock-market crash known as the South Sea Bubble (1720) and held it for over 20 years.

FAMILY PROBLEMS

George was a stolid, unimaginative man, despite his patronage of Handel, and out of his depth as a king in an alien country. His reputation was nevertheless somewhat sinister. He kept his wife a prisoner, with no access to her children, and was widely suspected of responsibility for the mysterious disappearance of her lover, Count Königsmarck, in 1694. Like his descendants, George I also quarrelled with his son and heir.

♛ George was anxious to keep abreast with the vogue for agricultural improvement. He once suggested planting turnips in St James's Park.

♛ George died of a stroke at Osnabrück after overeating on his way to Hanover.

George II

Born: Herrenhausen, Hanover, 30 Oct 1683. **Parents:** George I and Sophia Dorothea of Celle. **Ascended the Throne:** 11 June 1727. **Coronation:** Westminster Abbey, 11 Oct 1727. **Authority:** King of Great Britain and Ireland **Married:** Caroline, daughter of the Margrave of Brandenburg-Ansbach. **Children:** Four sons and five daughters; probably one or more illegitimate children. **Died:** Kensington Palace, 25 Oct 1760. **Buried:** Westminster Abbey.

George II spoke fluent English but with a strong accent and remained more German than English. He inherited a politically stable, economically prosperous and expanding kingdom, and a powerful minister, Walpole, whose survival, in spite of George's initial dislike, was largely due his close working relationship with the intelligent and worldly Queen Caroline (died 1737). Constitutional monarchy was becoming firmly established. Like his father, whom he detested, George hated his own eldest son, Frederick, Prince of Wales (1707-51), who died, to his father's ill-disguised glee, after being struck by a cricket ball.

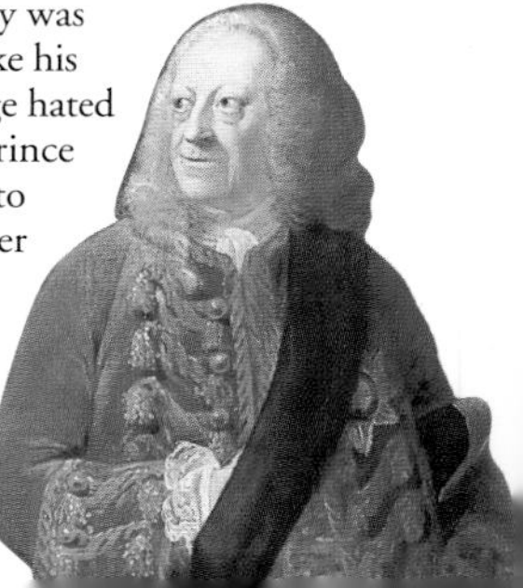

FOUNDATIONS OF EMPIRE

The contest with France for world-wide commercial

supremacy and the foundations of the British Empire were taking shape in George's reign, culminating in the military and naval victories of the Seven Years War (1756-63). British success was founded on the profits of agriculture and trade, which created investment capital, and increasing industrial production.

'DEVIL TAKE THE WHOLE ISLAND'

George resembled his father in appearance, in his liking for Handel – he originated the custom of standing up for the 'Hallelujah Chorus' in the *Messiah* – and in his preference for Hanover to England which, in a temper, he once consigned to the devil.

♛ One visible sign of economic growth was the building of turnpikes, the first national road system since the Romans.

♛ Dick Turpin, the famous highwayman, was arrested and executed in 1739.

♛ George II was the last British monarch to lead his troops on the battlefield, against the French at Dettingen in 1743, in the War of the Austrian Succession, and the last to be buried at Westminster.

♛ Government measures to reduce the appalling consumption of gin provoked a mob to harass the royal coach, shouting, 'No Gin, No King!'

Queen Caroline II

The Jacobite Risings

A constant, perhaps exaggerated, threat to the Hanoverian settlement was rebellion in favour of the legitimate, Stuart heir, James Edward ('The Old Pretender'). Jacobitism was strong in Scotland, especially the Highlands.

The Earl of Mar raised the clans in 1715 and captured Perth, but pro-Hanoverian clans commanded the north, and government forces built up rapidly in the south. By the time James Edward landed at Peterhead in December, the cause was already lost.

There were other disturbances before the last and most effective rising, led by James Edward's son, 'The Young Pretender', Prince Charles ('Bonnie Prince Charlie'), in 1745. Landing almost alone, he rallied the Jacobite clans by force of personality, conquered Scotland and invaded England. English support was, however, disappointing. At Derby Prince Charles was only 130 miles from London. Instinct told him to make a dash for the capital, but Lord George Murray and other lieutenants advised withdrawal. In spite of his conviction that retreat meant ruin, Charles agreed. In 1746 government forces under the Duke of Cumberland, a son of George II, annihilated

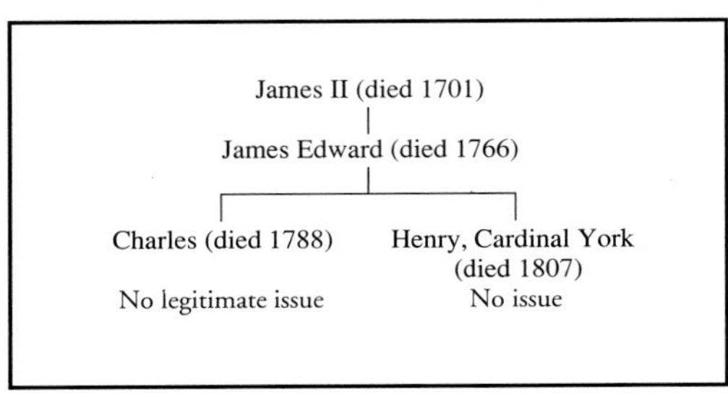

the Jacobites at Culloden, north-east of Inverness. Cumberland's brutality earned him the nickname 'The Butcher'. Charles escaped and, eventually, with the aid of Flora Macdonald, made his way to France.

After Culloden, the Highlands were brutally suppressed, the wearing of tartan forbidden, and every effort made to destroy the clan system and Gaelic culture.

George III

Born: Norfolk House, St James's Square, London, 4 June 1738. **Parents:** Frederick, Prince of Wales, and Augusta, daughter of the Duke of Saxe-Coburg-Gotha. **Ascended the Throne:** 25 Oct 1760. **Coronation:** Westminster Abbey, 22 Sept 1761. **Authority:** King of the United Kingdom of Great Britain and Ireland, Elector (later King) of Hanover. **Married:** Charlotte, daughter of the Duke of Mecklenburg-Strelitz. **Children:** Ten sons, including the future George IV and William IV, and six daughters. **Died:** Windsor, 29 Jan 1820. **Buried:** St George's Chapel, Windsor.

George III inherited the throne on a tide of domestic prosperity and martial victory. He prided himself on being British to the core and, with his amiable queen, echoed the tastes and prejudices of the rising middle class. Anxious to restore royal powers that had dwindled under his predecessors, he was practically the last monarch able to a large extent to choose his own ministers, and ended half a century of Whig dominance. His reign was clouded by periodic attacks of insanity, now known to have been symptoms of porphyria. His last attack (1810) proved permanent and his heir was proclaimed Prince Regent in February 1811.

George III and Lord Howe

AN IMPERIAL SETBACK

Growing disagreements between the British government and the North American colonists led to the American Declaration of Independence (1776) and successful rebellion. The king was widely blamed for the loss of the 13 colonies and is still popularly regarded as a wicked despot in the United States.

'FARMER GEORGE'

Dutiful, hard-working and genial, in spite of the revolutionary times, George made the monarchy more popular. He delighted in his nickname, born of his experimental farms in Windsor Park.

♛ George survived at least one assassination attempt, displaying commendable coolness under fire.
♛ The French Revolution and ensuing wars with France led to fears of invasion before the combined French and Spanish fleets were destroyed by Nelson at Trafalgar (1805).
♛ George fiercely, and successfully, opposed measures to restore the civil rights of Roman Catholics.

George IV

Born: St James's Palace, 12 Aug 1762. **Parents:** George III and Charlotte of Mecklenburg-Strelitz. **Ascended the Throne:** 29 Jan 1820. **Coronation:** Westminster Abbey, 19 July 1821. **Authority:** King of the United Kingdom of Great Britain and Ireland, King of Hanover. **Married:** Caroline, daughter of the Duke of Brunswick. **Children:** One daughter, Charlotte Augusta; two illegitimate children. **Died:** Windsor, 26 June 1830. **Buried:** St George's Chapel, Windsor.

According to Hanoverian custom, the Prince of Wales was on poor terms with his father whom he effectively succeeded when he became Prince Regent in 1811. An amiable, self-indulgent dandy with romantic illusions about himself (in old age he insisted he had fought at the Battle of Waterloo), he was the darling of society until he grew fat and over-extravagant. He married Caroline in 1795 only because parliament promised to pay his debts. Dislike was mutual, and though a bill to dissolve the marriage was dropped, Caroline was forcibly barred from the coronation. George was the first Hanoverian monarch to visit Ireland and Scotland (1822), wearing Highland

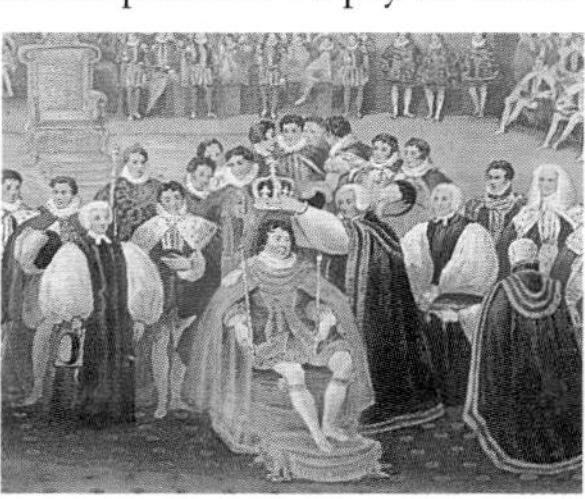

Coronation of George IV

dress – Stuart tartan – with some *élan* and winning over many by his charm.

O'CONNELL TAKES HIS SEAT

Elected in Co. Clare in 1828, Daniel O'Connell was debarred from parliament as a Roman Catholic. The Catholic Emancipation Act was forced through by the reforming home secretary, Sir Robert Peel, in 1829, with the backing of the Duke of Wellington but against the wishes of George IV.

ORIENTAL FANTASY

Although Parliament and others were appalled by the Prince Regent's extravagance, posterity has some reason to be grateful. Most notable of his artistic projects was the fairy-tale palace known as Brighton Pavilion.

- In 1785 George secretly, and illegally, married the most persistent of his many mistresses, Mrs Fitzherbert.
- When George first set eyes on the vulgar German princess he had agreed to marry, it is said he turned pale and asked for brandy.
- When the popular Princess Charlotte died in childbirth in 1817, George's brothers hastened to marry and produce heirs.

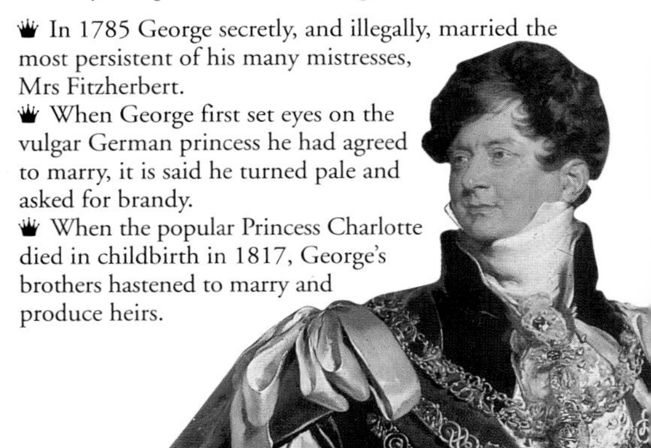

William IV

Born: Buckingham Palace, 21 August 1765.
Parents: George III and Charlotte of Mecklenburg-Strelitz.
Ascended the Throne: 26 June 1830.
Coronation: Westminster Abbey, 8 September 1831.
Authority: King of the United Kingdom of Great Britain and Ireland, King of Hanover.**Married:** Adelaide, daughter of the Duke of Saxe-Meinigen. **Children:** Four, none of whom survived infancy. **Died:** Windsor, 20 June 1837.
Buried: St George's Chapel, Windsor.

As Duke of Clarence and third son of George III, William IV had not expected to become king and always behaved more like a retired naval officer, which he was, than a monarch. He had a large family with an actress known as Mrs Jordan before marrying a young German princess in 1818. This attempt to produce an heir was unsuccessful, but Adelaide was a kindly stepmother to the brood of 'Fitzclarences'. Well-meaning but tactless, William lacked George IV's intelligence and charm but was more committed to duty and generally supported his ministers even when he disliked their policy.

PARLIAMENTARY REFORM

The political centrepiece of the reign was the Great Reform Act (1832), the first step on the road to universal franchise. William agreed to the request of Lord Grey, the prime minister, to dissolve the hostile parliament and put some pressure on the Lords to prevent the bill's defeat, though, worried by republicanism, he later got rid of Grey.

CROWNED 'ON THE CHEAP'

The coronation of William and Adelaide was in marked contrast to the gorgeous affair designed for himself by George IV. William insisted on economies, the usual banquet was cancelled, and Adelaide provided the jewels for her crown.

♛ The number of pubs called 'Queen Adelaide' testifies to the popularity of William IV's sweet-natured queen.
♛ Six Dorset farm labourers, the 'Tolpuddle Martyrs', were transported to Australia in 1834 for conspiring to form a trade union.
♛ In 1836 the South African Boers set out on the 'Great Trek' to escape British control of Cape Colony.

Victoria

Born: Kensington Palace, 24 May 1819. **Parents:** Edward, Duke of Kent, and Victoria, daughter of the Duke of Saxe-Coburg-Saalfeld. **Ascended the Throne:** 20 June 1837. **Coronation:** Westminster Abbey, 28 June 1838. **Authority:** Queen of the United Kingdom of Great Britain and Ireland, Empress of India (from 1 May 1876). **Married:** Albert, son of the Duke of Saxe-Coburg-Gotha. **Children:** Four sons, including the future Edward VII, and five daughters. **Died:** Osborne, Isle of Wight, 22 Jan 1901. **Buried:** Frogmore, Windsor.

During Victoria's reign, the longest of any British monarch, the Crown lost virtually all executive power but became a potent national symbol, representing constancy and stability in changing times. Victoria's sense of duty and her sympathy with the opinions of the dominant middle class were augmented by Albert, created Prince Consort in 1857, a man of ability and diligence. His early death (1861) plunged her into prolonged and secluded mourning, temporarily losing popularity and stimulating republicanism. While personifying strict Victorian morality, Victoria caused consternation at court by her fondness for certain servants, such as her devious Indian secretary, 'the Munshi', and the forthright Highland gillie, John Brown.

PROGRESS AND ORDER

In spite of the tensions caused by rapid social and economic change, for most of Victoria's reign Britain was extraordinarily orderly. Political violence was rare after 1848 when the Chartists, demanding political reforms, staged a protest in London. Crime actually decreased. Education and democracy made large advances. Social reforms ameliorated the hardships of industrialization.

THE ROYAL 'WE'

Queen Victoria has been mocked for saying 'we' when referring to herself. It resulted from her reluctance to dissociate Prince Albert, even after his death, from her own views. In later life she denied making the famous remark, 'We are not amused.'

♛ Victoria could not conceal her preferences among her ministers, adoring the Tory Benjamin Disraeli (prime minister 1868, 1874-80), loathing the Liberal William Gladstone (prime minister 1868-74, 1880-5, 1886, 1892-4).

♛ At Victoria's accession, the fastest transport was the horse. By her death, trains were being challenged by motor cars.

♛ At Victoria's death, few of her subjects were old enough to remember life under another monarch.

The Role of Monarchy under Victoria

The role of the monarchy as we know it today was established in the reign of Victoria (1837-1901). While losing almost all executive power, the monarchy became an office of great prestige, with the monarch representing all the people – not merely an aristocratic clique. A grand symbol of the national heritage, Victoria was also an ordinary person with whom her subjects could identify, and this dual role has been played, with varying success, by her successors. Victoria herself was largely responsible for the changed image of monarchy, although not by design: she would not have willingly surrendered any royal powers if she could have prevented it.

Victoria was not always popular. She lost favour by her withdrawal from public life after the death of Prince Albert. (In

1870, when there was an outburst of republicanism, she performed only one public ceremony, opening Blackfriars Bridge.) But in later years, lured out of retirement by Benjamin Disraeli (below), she became more popular than ever.

The monarchy lost power under Victoria but not grandeur. Her Diamond Jubilee (1897) was a dazzling imperial pageant, monarchy at its most theatrical, but the centre of attention was a dumpy little old lady in black.

Home Rule

Desire for Home Rule – limited autonomy under the Crown – was manifest in Scotland, Wales and especially Ireland during the 19th century.

In Ireland the majority of the population were Roman Catholics and most of the land was held by a Protestant élite. The catastrophe of the Great Famine (one million died, two million emigrated) in 1845-6 provoked violent outbreaks by a

Nelson's Pillar, Dublin

few revolutionaries. Reforms, such as disestablishment of the Anglican Church and the Land Act of 1870, failed to quell the movement for Home Rule, which parliament conceded in 1914. Before it came into effect, the Easter Rising in Dublin (1916) and its aftermath led to the treaty of 1921 which created the Irish Free State. To appease Protestant Ulster, six northern counties remained within the United Kingdom.

In Scotland and Wales the majority of the population were less alienated by religious discrimination and absentee landlords. Both countries, unlike Ireland, played important parts in Britain's industrial and imperial growth. Economic decline and the disappearance of the British Empire after the Second World War encouraged demands for internal autonomy and, in Scotland particularly, for independence, though nationalist parties remained in the minority.

The Irish Home Rule Maze

Parliament and Sovereignty

A powerful theme of British history is the struggle for sovereignty between the Crown and parliament, specifically the House of Commons. Parliament developed from the medieval Curia Regis, or 'royal council', an advisory assembly of magnates. From the 13th century representatives of humbler classes were sometimes summoned. They met separately, and the two bodies developed into Lords and Commons.

The gradual ascendancy of parliament began with its control of revenue. As feudal dues faded, parliamentary grants of

taxation became vital to the Crown. The victory of parliament in the English Civil Wars put paid to the doctrine of the divine right of kings, and parliamentary sovereignty was confirmed by the Glorious Revolution of 1688 and the ensuing Bill of Rights.

The milestones on the road to the supremacy of parliament were less important than the gradual erosion of royal power as a result of less dramatic developments, such as the growth of cabinet government under the early Hanoverians and the extension of the franchise by the Reform Acts of the 19th century.

Charles I had recognized virtually no limits to the power of the Crown. By Victoria's reign, these powers had practically vanished. According to the constitutional expert Walter Bagehot (1826-77), the monarch could only advise, not give orders.

The British Empire

Queen Victoria reigned over a worldwide empire, which at its peak in 1914 included about one-fifth of the world's land surface and one-quarter of its population. Besides the areas coloured red on the map opposite , Britain also exercised a powerful influence over large areas in Latin America and the Far East.

The greatest single motive for Britain's imperial expansion was the doctrine of free trade, which the British were prepared, if necessary, to extend by force. There were other motives. Australia, for example, was first colonized as a convict settlement, while some African colonies were acquired for strategic reasons, specifically to frustrate the French.

Apart from the self-governing dominions (Canada from 1867, Australia from 1901 and South Africa from 1910), the core of the empire – the greatest jewel in Victoria's crown – was India. Victoria, jealous of the imperial titles of fellow-monarchs in Europe, was proclaimed Empress of India in 1876.

The colonies provided Britain with opportunities for its rapidly expanding population, though often at the expense of indigenous peoples.

The conquest of India was one of the pinnacles of Queen Victoria's reign

The British Empire in 1886

Edward VII

Born: Buckingham Palace, 9 November 1841.
Parents: Victoria and Albert.
Ascended the Throne: 22 January 1901.
Coronation: Westminster Abbey, 9 August 1902.
Authority: King of Great Britain and Ireland and of British Dominions overseas, Emperor of India.
Married: Alexandra, daughter of Christian IX of Denmark.
Children: Three sons, including the future George V, and three daughters. **Died:** Buckingham Palace, 6 May 1910.
Buried: St George's Chapel, Windsor.

In character the sybaritic Edward VII was most unlike his parents. He enjoyed social and sporting life, and as Prince of Wales, was linked with several scandals which gave his mother an excuse for excluding him from royal duties other than purely ceremonial ones. He visited North America and India among other state visits, and his genial nature and love of ceremonial occasions – the theatrical side of monarchy – made him widely popular. His personal influence helped to cement the Entente Cordiale between Britain and France, though he was less successful with his nephew, the German Emperor, whom he disliked.

SOCIAL WELFARE

Following the victory of the Liberals in the general election of 1906, the first steps were taken towards the creation of a welfare state with the introduction of national insurance and old-age pensions. The Liberal programme led to a constitutional crisis in 1910 when the Tory-dominated Lords refused to pass Liberal legislation.

NO VOTES FOR WOMEN

Like most monarchs Edward was a conservative, to whom democracy was still anathema. He opposed granting women the vote, but that view was widely shared among his subjects.

♛ Edward VII was the only monarch of the dynasty of Saxe-Coburg-Gotha, or House of Wettin.
♛ Edward ate five large meals every day, dinner usually running to at least ten courses.
♛ Though his reign was short, Edward gave his name to an age which in many ways he personified.

George V

Born: Marlborough House, London, 3 June 1865. **Parents:** Edward VII and Alexandra. **Ascended the Throne:** 6 May 1910. **Coronation:** Westminster Abbey, 22 June 1911. **Authority:** King of Great Britain and Ireland and British Dominions overseas, Emperor of India. **Married:** Mary, daughter of the Duke of Teck. **Children:** Five sons, including the future Edward VIII and George VI, and one daughter. **Died:** Sandringham, Norfolk, 20 January 1936. **Buried:** St George's Chapel, Windsor.

Like William IV, George V was a younger son who was not expected to become king and therefore served in the navy. He retained a certain informality, which suited his simple, straightforward character and reinforced his popularity. He proved to be the model modern monarch by carrying out his royal responsibilities with exemplary dedication while maintaining an irreproachable personal and family life, although, like many predecessors, he fell out with the Prince of Wales. On several occasions he showed that the monarch's right to advise could be significant, especially in pressing for an end to the violence in Ireland, 1919-21.

THE FIRST WORLD WAR

A century of relative peace in Europe ended in 1914 with the outbreak of the First World War, in which nearly one million British subjects were killed and two-and-a-half million wounded. Among other casualties were several European monarchies, but in spite of tumultuous social and political changes the British monarchy emerged intact.

HIS FAVOURITE OPERA

Unlike his artistically inclined queen, George V cared nothing for the arts and admitted to uncultured tastes. He once said that his favourite opera was *La Bohème,* because it was 'the shortest'.

♛ George V's elder brother, Clarence, who died in 1892, had a doubtful reputation. George was a much steadier character.

♛ King George and Queen Mary were crowned emperor and empress of India in a fabulous ceremony at the Delhi Durbar on 12 December 1911.

♛ George V was the first monarch to make a Christmas Day broadcast to the nation.

Edward VIII

Born: White Lodge, Richmond, Surrey, 23 June 1894. **Parents:** George V and Mary of Teck. **Ascended the Throne:** 20 Jan 1936. **Coronation:** Never crowned. **Authority:** King of Great Britain and Ireland and British Dominions overseas, Emperor of India. **Married:** Mrs Wallis Simpson. **Children:** None. **Died:** Paris, 28 May 1972. **Buried:** Frogmore, Windsor.

George V's eldest son, known as David to the royal family, appeared an ideal Prince of Wales. Handsome and charming, a great success on public occasions and foreign visits, he also expressed sympathy with the unemployed. His predilection for fashionable society and for older, married women worried his father, and he resented George V's refusal to allow him to serve at the front during the world war. On succeeding to the throne, his plan to marry Mrs Wallis Simpson, an American divorcee, provoked a constitutional crisis ending in his abdication after 11 months. Created Duke of Windsor (1937), he lived the rest of his life abroad.

THE PRINCE AND THE NAZIS

Edward's political naïvety led him to make unwisely approving remarks about the Nazi regime in Germany during the 1930s. Suggestions that he was a traitor to Britain during the Second World War, during which he served as Governor of the Bahamas (1940-45), and hoped to regain the throne after a German invasion, seem to be exaggerated.

WALLIS UNDER WRAPS

The British press agreed to conceal the king's liaison with Mrs Simpson. Pictures showing them together were censored, but as they appeared in foreign papers, the affair gradually became public knowledge.

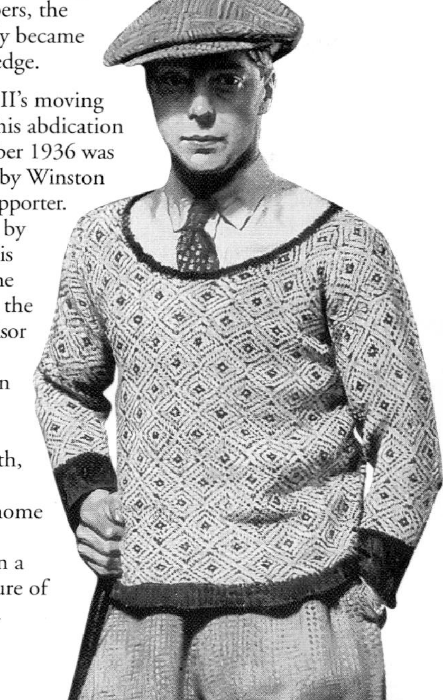

- Edward VIII's moving broadcast on his abdication on 11 December 1936 was partly drafted by Winston Churchill, a supporter.
- Influenced by his wife and his affinity with the United States, the Duke of Windsor spoke with a slight American accent.
- Ten days before his death, the Duke was visited at his home by his niece, Elizabeth II, in a symbolic gesture of reconciliation.

The Abdication Crisis

The woman with whom the Prince of Wales fell in love was Mrs Wallis Simpson (1896-1986), an American, then married and divorced from a previous husband. The main constitutional objection was that, as head of the Church of England, the King could hardly marry a twice-divorced woman. That she was also a commoner, a member of a distinctly 'fast' social set, and widely disliked, did not help. The king, once accused by his mother of the inability to see any point of view but his own, wanted both to be king and to marry Mrs Simpson. The prime minister, Stanley Baldwin, convinced him that this was impossible, and Edward chose Mrs Simpson rather than the Crown. They were married in France after her divorce was finalized in June 1937.

The ex-king was created Duke of Windsor but with the – to him – mortifying qualification that the title extended to him alone. His wife was thus, whatever the legality of such a condition, denied the title of 'Her Royal Highness', an indication of the bitterness felt towards her in the royal family.

Edward had many supporters, including Winston Churchill. However, Churchill later held that it all turned out for the best, since Edward's younger brother proved himself to be an ideal monarch, and his wife an ideal consort.

George VI

Born: Sandringham, Norfolk, 14 Dec 1895. **Parents:** George V and Mary of Teck. **Ascended the Throne:** 11 Dec 1936. **Coronation:** Westminster Abbey, 12 May 1937. **Authority:** King of the United Kingdom of Great Britain and Northern Ireland and British Dominions overseas, Emperor of India (until 1947). **Married:** Elizabeth Bowes-Lyon, daughter of the Earl of Strathmore and Kinghorne. **Children:** Two daughters, the future Elizabeth II and Margaret. **Died:** Sandringham, 6 February 1952. **Buried:** St George's Chapel, Windsor.

In the tradition of younger sons, Prince Albert ('Bertie': he adopted the name George in tribute to his father) served in the navy and was later created Duke of York. Shy, with a stammer that made public speaking an ordeal, he was appalled to be summoned to the throne at three weeks' notice after Edward VIII's abdication, and so was his wife. With all his father's transparent decency and devotion to duty, and an underlying humility (despite occasional flashes of Hanoverian temper), he proved an ideal figurehead during the Second World War. Refusing to leave London, the king and queen did wonders for Londoners' morale during the Blitz, narrowly escaping death themselves when Buckingham Palace was hit.

THE SECOND WORLD WAR

During nearly half of George VI's reign Britain was at war, and in 1940-1, in spite of support from the Empire, especially the Dominions, appeared to be in danger of defeat. The great changes that war brought, or expedited, included the relinquishing of empire. George VI lost his imperial title with Indian independence in 1947 and became head of the (British) Commonwealth in 1949.

DOUBTS ABOUT PRIME MINISTERS

When Neville Chamberlain resigned as prime minister in May 1940, the king would have preferred Lord Halifax to Churchill, who had a reputation as a political maverick. However, their relationship ripened into mutual admiration and friendship. In 1945 he was similarly cool, at first, towards the Labour prime minister, Clement Attlee.

- George VI restored the popularity of the monarchy, damaged by the Abdication Crisis.
- In the 1930s the king's stammer was much reduced, though not quite cured, by an Australian speech therapist.
- From childhood George VI suffered frequent illness. He died not long after an operation for lung cancer.

Elizabeth II

Born: 17 Bruton Street, London, 21 April 1926.
Parents: George VI and Elizabeth Bowes-Lyon.
Ascended the Throne: 6 February 1952.
Coronation: Westminster Abbey, 2 June 1953.
Authority: Queen of the United Kingdom of Great Britain and Northern Ireland, Head of the Commonwealth. **Married:** Philip, son of Prince Andrew of Greece. **Children:** Three sons, Charles, Andrew and Edward, and one daughter, Anne.

The Queen inherited her father's devotion to duty, simple tastes and fondness for family life. She became officially engaged to Prince Philip, an officer in the Royal Navy, in 1947, and he was created Duke of Edinburgh. He brought vigour and informality to the royal family, at the cost of occasional breaches of tact. The Queen was criticized for being out of touch and priggish during an anti-monarchical spasm around 1957, but otherwise commanded widespread respect and affection. The prestige of royalty, however, was seriously damaged by the much publicized failures of the marriages of, first, her sister, Princess Margaret, and later of three of her children, including the Prince of Wales.

THE COMMONWEALTH

During the reign of Elizabeth II, Britain has relinquished the last remnants of empire apart from a few minor territories. A shadow of empire remained in the Commonwealth, an association of Anglophone nations, some of which acknowledged the Queen as ceremonial head of state, but the Suez affair (1956) demonstrated Britain's international weakness.

ROYAL SOAP OPERA

Elizabeth II has played her dual role as representative of the people and symbol of the nation faultlessly, and has revealed more of her private life to the public than any predecessor. But the activities of some members of the royal family, publicized by ruthless news media, has threatened to turn the royal family into a source of cheap entertainment.

- To defuse criticism of royal wealth, the Queen agreed to pay taxes on her private income.
- Since there is no tradition of abdication, in the event of the Queen's incapacity the probable solution would be a regency.
- The Queen has made more state visits than any previous monarch.

COMPENDIUM 

The Crown Jewels

The English regalia, popularly known as the 'Crown Jewels', are on public view in the Tower of London. Most of them are used only at a coronation. They are not as venerable as people often suppose, as the original regalia were sold and melted down during the Interregnum.

The Crown used for the coronation is St Edward's Crown, made in 1661 and modelled on its lost predecessor. **The Imperial State Crown,** made for Victoria's coronation in 1838, is worn at state openings of parliament. Though lighter than St Edward's Crown, George V complained that it gave him a headache. The **Imperial Crown of India** was made for George V's coronation at Delhi.

The Sceptre, symbol of royal power and justice, is held in the sovereign's right hand. Originally made in 1660, it contains a massive heart-shaped gem cut from the Cullinan Diamond.

The Orb, held in the left hand, symbolizes Christian dominion over the Earth, though it was a symbol of royal authority in pre-Christian times.

The Ampulla contains the oil with which the sovereign is annointed – the most important and oldest rite of the coronation ceremony. The oil is poured into the **Spoon.** These objects escaped dispersion in the 17th century. The

ampulla was probably made for the coronation of Henry IV in 1399. The spoon was made over a century earlier.

The Armils are bracelets that have been worn at coronations since the Middle Ages. Their origin is obscure. They seem to date back to a custom of early German kings, if not to the Jewish kings of the Old Testament.

The Sword of State, symbol of authority and of the rite of investiture, is carried before the monarch at the coronation. There are five swords among the English regalia.

The Scottish Crown Jewels, the 'Honours of Scotland', which can be seen in Edinburgh Castle, were last used in 1651. The Crown dates from 1488, though it was remodelled in 1540. The Sceptre and the Sword of State were gifts to James IV from the Pope in 1494 and 1507 respectively.

The Irish Crown Jewels were stolen from Dublin Castle and have never been traced.

Medieval kings had several crowns: Edward II had at least ten. One advantage was that in times of financial difficulty, they could be sold or pawned.

Underneath the Crown, the sovereign wears a velvet cap known as the Cap of Maintenance. What this signifies is a mystery.

The Imperial State Crown is probably the most valuable in the world. It contains a sapphire from the coronation ring of Edward the Confessor, a ruby that belonged to the Black Prince, a sapphire from the crown of Charles II and one of the gems cut from the Cullinan diamond.

What is probably still the most famous diamond in the world, and at one time the largest, the Indian Koh-i-Noor, first documented in India in 1304, was presented to Queen Victoria by the East India Company in 1849. It was incorporated in the crown made for Elizabeth, now the Queen Mother, for the coronation of George VI in 1937.

English monarchs since Edward II have been crowned in the **Coronation Chair** in Westminster Abbey. It was made for Edward I about 1300, with a space to house the **Stone of Scone,** captured in 1298, on which Scottish kings were consecrated.

What is believed to be the stone on which West Saxon kings, possibly including Alfred the Great, were consecrated can still be seen in Kingston-upon-Thames.

Nicknames

Edmund I (d.946): the Elder, the Magnificent
Edwy (d.959): the Fair
Edgar (d.975): the Peaceful
Sweyn (d.1014): Forkbeard
Harold I (d.1040): Harefoot
Ethelred (d.1016): the Redeless, the Unready
Edmund II (d.1016): Ironside
Edward (d.1066): the Confessor
Malcolm III: *Ceann mor* ('Canmore')
Alexander I: the Fierce
David I: the Saint
Malcolm IV: the Maiden
William I: the Lion
James I: the Scottish Orpheus
James II: Fiery Face
James V: the King o' the Commons
William I: the Bastard, the Conqueror*
William II: Rufus
Henry I: Beauclerk
Henry II: Curtmantle
Richard I: *Coeur de Lion* ('the Lionheart')
John: Lackland
Edward I: Longshanks, Hammer of the Scots*
Henry V: the English Alexander*
Henry VI: the Martyr King
Edward IV: the Robber
Richard III: Crouchback
Henry VIII: Bluff King Hal
Mary I: Bloody Mary
Elizabeth: Gloriana, Good Queen Bess
James I: the British Solomon, the Wisest Fool in Christendom
Charles I: the Martyr
Charles II: Old Rowley, the Merry Monarch
James II: the Popish Duke
William III: Dutch Billy
Anne: Brandy Nan
George III: Farmer George
George IV: Prinny, First Gentleman of Europe
William IV: the Sailor King, Silly Billy
Victoria: the Widow of Windsor, Grandmother of Europe
Edward VII: Teddy, the Peacemaker
George V: the Sailor King

*posthumously

Genealogical Chart

THE HOUSE OF MACALPIN 834-1034

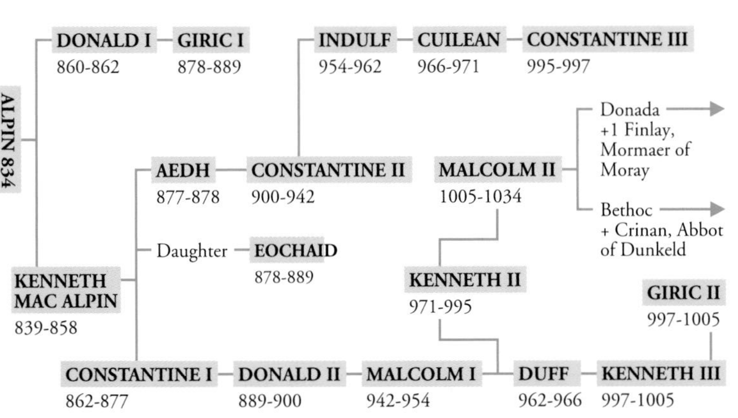

THE HOUSE OF DUNKELD 1034-1371

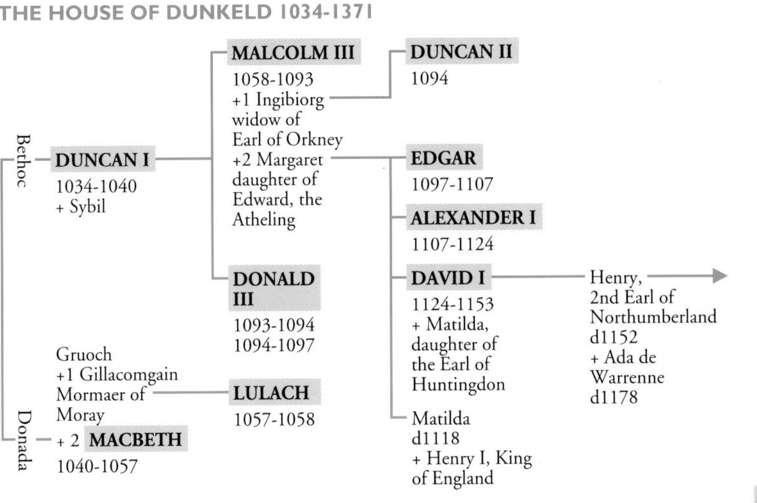

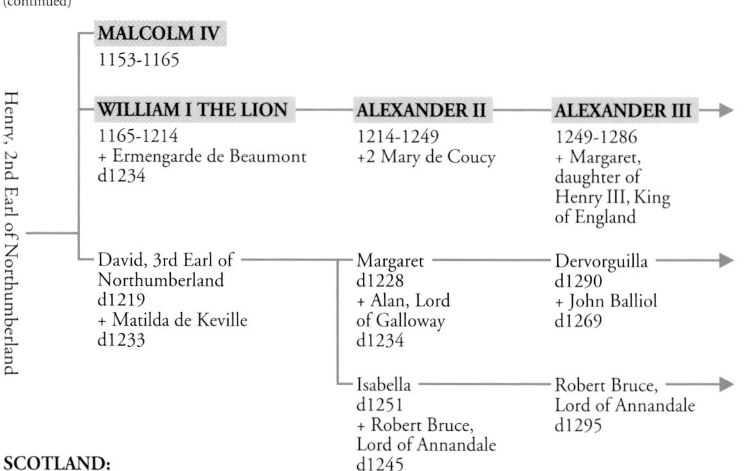
THE HOUSE OF DUNKELD 1058-1371
(continued)
Henry, 2nd Earl of Northumberland
MALCOLM IV
1153-1165
WILLIAM I THE LION
1165-1214
+ Ermengarde de Beaumont
d1234
ALEXANDER II
1214-1249
+2 Mary de Coucy
ALEXANDER III
1249-1286
+ Margaret,
daughter of
Henry III, King
of England
David, 3rd Earl of
Northumberland
d1219
+ Matilda de Keville
d1233
Margaret
d1228
+ Alan, Lord
of Galloway
d1234
Dervorguilla
d1290
+ John Balliol
d1269
Isabella
d1251
+ Robert Bruce,
Lord of Annandale
d1245
Robert Bruce,
Lord of Annandale
d1295
SCOTLAND:

Margaret
d1283
+ Erik II, King of Norway
d1299

MARGARET
1286-1290

THE HOUSE OF BALLIOL

JOHN BALLIOL
1292-1296
+ Isabella de Warenne

EDWARD BALLIOL
1332-1341

THE HOUSE OF BRUCE

Robert Bruce
d1304
+ Marjorie,
Countess of Carrick

ROBERT I
1306-1329
+2 Elizabeth de Burgh
d1327

Marjorie Bruce
d1316
+ Walter, the Steward
d1326

DAVID II
1329-1371
+ Joan, daughter of
Edward II, King of England

THE HOUSE OF STEWART 1371-1567

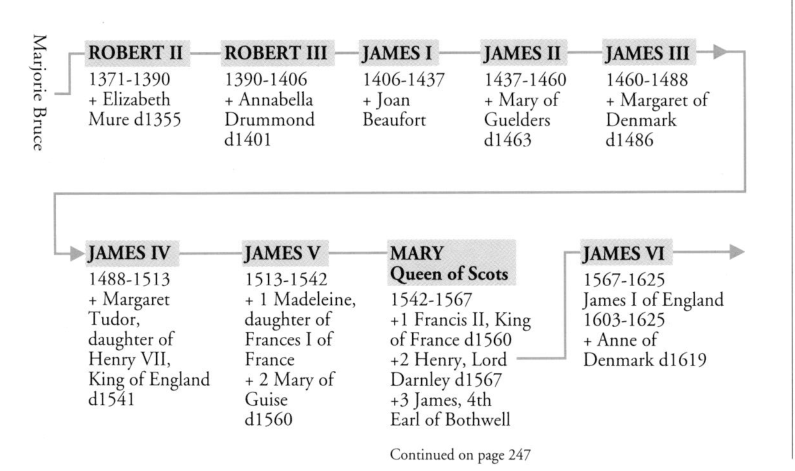

Continued on page 247

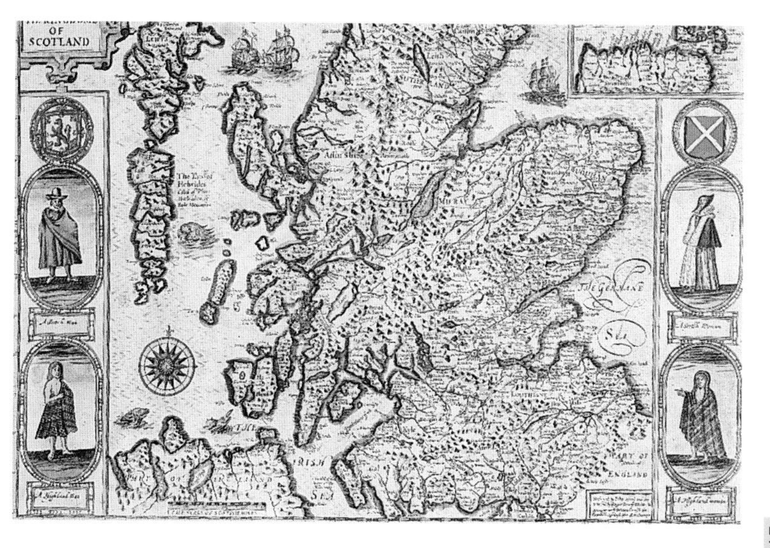
OF
SCOTLAND
THE GERMANE
IRISH
SEA
PART OF
ENGLAND

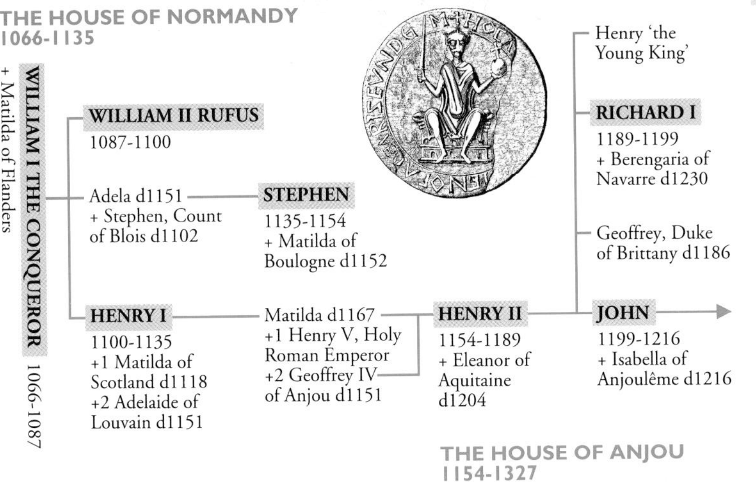
THE HOUSE OF NORMANDY
1066-1135
WILLIAM I THE CONQUEROR 1066-1087
+ Matilda of Flanders
WILLIAM II RUFUS
1087-1100
Adela d1151
+ Stephen, Count of Blois d1102
STEPHEN
1135-1154
+ Matilda of Boulogne d1152
HENRY I
1100-1135
+1 Matilda of Scotland d1118
+2 Adelaide of Louvain d1151
Matilda d1167
+1 Henry V, Holy Roman Emperor
+2 Geoffrey IV of Anjou d1151
HENRY II
1154-1189
+ Eleanor of Aquitaine d1204
Henry 'the Young King'
RICHARD I
1189-1199
+ Berengaria of Navarre d1230
Geoffrey, Duke of Brittany d1186
JOHN
1199-1216
+ Isabella of Anjoulême d1216
THE HOUSE OF ANJOU
1154-1327

JOHN

HENRY III
1216-1272
+ Eleanor of
Provence d1291

EDWARD I
1272-1307
+ Eleanor of
Castile d1290

EDWARD II
1307-1327
+ Isabella of
France d1358

EDWARD III
1327-1377
+ Philippa of
Hainault d1369

Eleanor d1275
+2 Simon de
Montfort, Earl
of Leicester d1265

EDWARD III

Edward, Prince of Wales,
The Black Prince d1376
+ Joan of Kent d1385

RICHARD II
1377-1399
+1 Anne of Bohemia d1394
+2 Isabella of France d1409

Lionel 1st Duke of Clarence d1368

THE HOUSE OF LANCASTER 1399-1471

John of Gaunt, 2nd Duke of
Lancaster d1399
+1 Blanche of Lancaster d1369

HENRY IV
1399-1413
+1 Mary of Bohun d1394
+2 Joan of Navarre d1437

+2 Catherine of Castile
+3 Catherine Swynford d1403

John Beaufort,
1st Marquess of Dorset d1410
+ Margaret Holland

THE HOUSE OF YORK 1461-1485

Edmund of Langley,
1st Duke of York d1482
+1 Isabella of Castile
+2 Joan Holland

Richard, 4th
Earl of Cambridge d1415
+ Anne Mortimer, great grand daughter
of Lionel, 1st Duke of Clarence

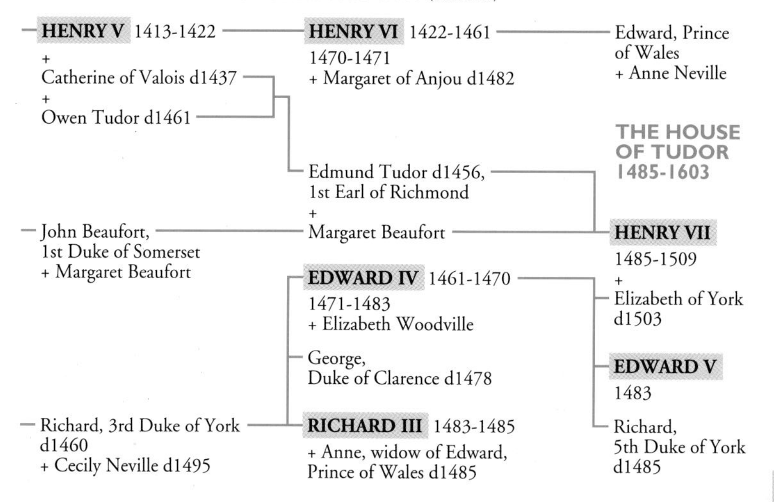
THE HOUSE OF LANCASTER 1399-1471 (continued)
HENRY V 1413-1422
+
Catherine of Valois d1437
+
Owen Tudor d1461
HENRY VI 1422-1461
1470-1471
+ Margaret of Anjou d1482
Edward, Prince
of Wales
+ Anne Neville
THE HOUSE
OF TUDOR
1485-1603
Edmund Tudor d1456,
1st Earl of Richmond
+
Margaret Beaufort
John Beaufort,
1st Duke of Somerset
+ Margaret Beaufort
HENRY VII
1485-1509
+
Elizabeth of York
d1503
EDWARD IV 1461-1470
1471-1483
+ Elizabeth Woodville
George,
Duke of Clarence d1478
EDWARD V
1483
Richard, 3rd Duke of York
d1460
+ Cecily Neville d1495
RICHARD III 1483-1485
+ Anne, widow of Edward,
Prince of Wales d1485
Richard,
5th Duke of York
d1485

THE HOUSE OF TUDOR 1485-1603 (continued)

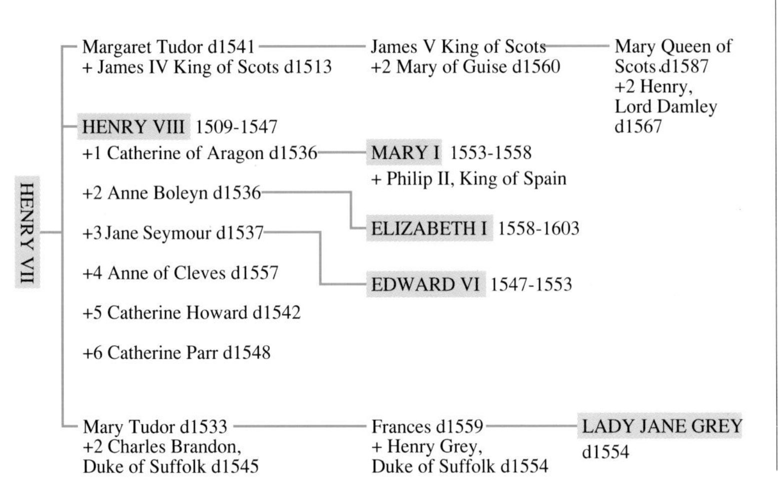
HENRY VII
Margaret Tudor d1541
+ James IV King of Scots d1513
James V King of Scots
+2 Mary of Guise d1560
Mary Queen of Scots d1587
+2 Henry, Lord Damley d1567
HENRY VIII 1509-1547
+1 Catherine of Aragon d1536
MARY I 1553-1558
+ Philip II, King of Spain
+2 Anne Boleyn d1536
ELIZABETH I 1558-1603
+3 Jane Seymour d1537
EDWARD VI 1547-1553
+4 Anne of Cleves d1557
+5 Catherine Howard d1542
+6 Catherine Parr d1548
Mary Tudor d1533
+2 Charles Brandon, Duke of Suffolk d1545
Frances d1559
+ Henry Grey, Duke of Suffolk d1554
LADY JANE GREY
d1554

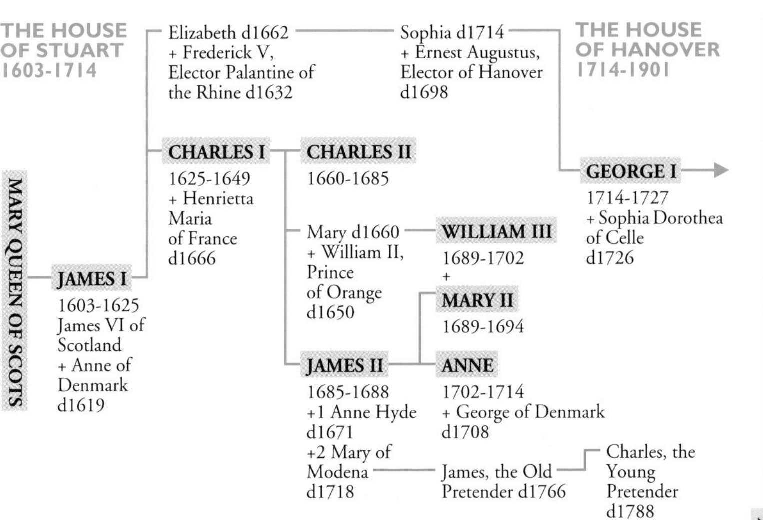
THE HOUSE OF STUART 1603-1714
MARY QUEEN OF SCOTS
JAMES I
1603-1625 James VI of Scotland + Anne of Denmark d1619
Elizabeth d1662 + Frederick V, Elector Palantine of the Rhine d1632
CHARLES I
1625-1649 + Henrietta Maria of France d1666
CHARLES II
1660-1685
Mary d1660 + William II, Prince of Orange d1650
JAMES II
1685-1688 +1 Anne Hyde d1671 +2 Mary of Modena d1718
Sophia d1714 + Ernest Augustus, Elector of Hanover d1698
WILLIAM III
1689-1702
+
MARY II
1689-1694
ANNE
1702-1714 + George of Denmark d1708
James, the Old Pretender d1766
Charles, the Young Pretender d1788
THE HOUSE OF HANOVER 1714-1901
GEORGE I
1714-1727 + Sophia Dorothea of Celle d1726

THE HOUSE OF HANOVER 1714-1901

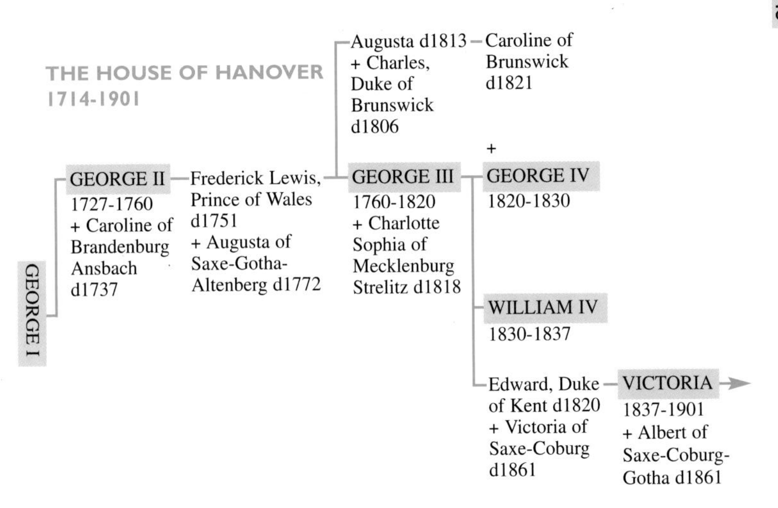

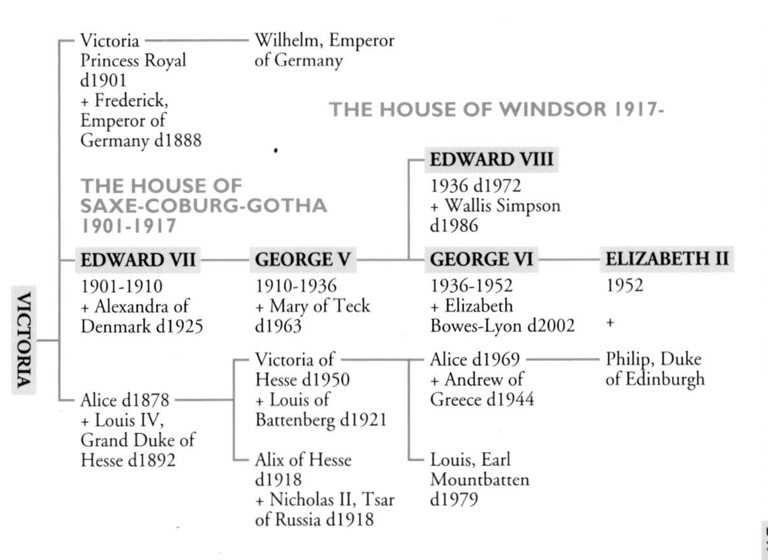
VICTORIA
Victoria
Princess Royal
d1901
+ Frederick,
Emperor of
Germany d1888
Wilhelm, Emperor
of Germany
THE HOUSE OF WINDSOR 1917-
THE HOUSE OF
SAXE-COBURG-GOTHA
1901-1917
EDWARD VIII
1936 d1972
+ Wallis Simpson
d1986
EDWARD VII
1901-1910
+ Alexandra of
Denmark d1925
GEORGE V
1910-1936
+ Mary of Teck
d1963
GEORGE VI
1936-1952
+ Elizabeth
Bowes-Lyon d2002
ELIZABETH II
1952
+
Philip, Duke
of Edinburgh
Alice d1878
+ Louis IV,
Grand Duke of
Hesse d1892
Victoria of
Hesse d1950
+ Louis of
Battenberg d1921
Alix of Hesse
d1918
+ Nicholas II, Tsar
of Russia d1918
Alice d1969
+ Andrew of
Greece d1944
Louis, Earl
Mountbatten
d1979

Illustration Notes

All pictures were supplied courtesy of the Bridgeman Art Library, London (coded BAL throughout), except for those individually credited below. Picture numbers precede the illustration reference.

5 Queen Elizabeth II, 1954, by Pietro Annigoni (Fishmongers' Hall, London). BAL. 6 The Prince of Wales (Henry V) Pays Homage to King Charles VI (Bibliotheque de Toulouse). BAL. 7 The Royal Oak Tree (English Revolution), Engraving, 1651 (British Library). BAL. 8 Seal of Edward The Confessor (Private Collection). 11 The Ancient Britons Acknowledging the Supremacy of Caesar (Private Collection). 13 Julius Caesar (Private Collection). 14 Selection of Celtic a Coins Found at Erisell, Suffolk (British Museum, London). BAL. 15 Boadicea, Queen of the Iceni, designed by C. H. S., (Private Collection). BAL. 17 Head of Caesar Augustus, (Location unknown). BAL. 21 King Arthur in Combat (Bibliothèque Nationale, Paris). BAL. 22 Arms and Costume of a Saxon Military Chief (Private Collection). Courtesy of Hulton Deutsch Collection Limited, London. 23 Sestertius of Hadrian from Newcastle Upon Tyne. Shows Bust of Hadrian as a Victorius Commander (Museum of Antiquities, Newcastle Upon Tyne). BAL. 24 Arms and Costume of an Anglo-Saxon King and Armour-Bearer (Private Collection). Courtesy of Hulton Deutsch Collection Limited, London. 27 Macbeth Instructing the Murderers Employed to Kill Banquo, by George Cattermole (Victoria & Albert Museum, London). BAL. 29 King Malcolm III. Courtesy of Hulton Deutsch Collection Limited, London. 31 King Edgar Seated between Saint Dunstan and Saint Ethelwold, Christchurch, from Regularis Concorda and Rule of Saint Benedict (British Library, London). BAL. 32 King David I of Scotland. Courtesy of Hulton Deutsch Collection Limited, London. 34 David I, King of Scotland from 1124, and his successor, King Malcolm IV, King from 1153. Courtesy of Hulton Deutsch Collection Ltd. 35 William 'The Lion'. Courtesy of Hulton Deutsch Collection Ltd. 37 King Alexander III of Scotland. Courtesy of Hulton Deutsch Collection Limited, London. 38 Robert Bruce, from the Seton Armorial. Courtesy of The National Library of Scotland. 40 Bruce's Tomb, Dunfermline Abbey. Courtesy of The Still Moving Picture Company. 41 Bruce at Bannockburn. (Private Collection). 42 David II, King of Scotland from 1329. Courtesy of Hulton Deutsch Collection Limited, London. 43 Edward 'Lion', King of Scotland, Courtesy of Hulton Deutsch Collection Limited, London. 45 James I of Scotland, by Anonymous (Scottish National Portrait Gallery, Edinburgh). BAL. 46 James II of Scotland, by Anonymous (Scottish National Portrait Gallery, Edinburgh). BAL. 47 James III of Scotland by Anonymous (Scottish National Portrait Gallery, Edinburgh). BAL. 50 James V of Scotland, by Anonymous (Scottish National Portrait Gallery, Edinburgh). BAL. 51 Mary, Queen of Scots, by William 'Senior' Bone (Wallace Collection, London). BAL. 52 James VI of Scotland and I of England and Ireland, by Adam de Colone (Scottish National Portrait Gallery, Edinburgh). BAL. 53 The Execution of Mary, Queen of Scots, by Anonymous (Scottish National Portrait Gallery, Edinburgh). BAL. 54 King Cormac and Fair Eithne, by Thomas Wright. Courtesy of Hulton Deutsch Collection Limited, London. 55 Irish King, Brian Boru is killed by a Viking Soldier, in a Surprise Attack Whilst he is Praying in his Tent. Courtesy of Hulton Deutsch Collection Limited, London. 57 Richard II, Kings of England from 1377 to 1399, Knighting the Four Kings of Ireland (O'Neill, O'Connor, O'Brien and MacMorrogh), in Dublin. Courtesy of Hulton Deutsch Collection Limited, London. 58 The Tara Brooch (National Museum of Ireland, Dublin). Courtesy of Visual Arts Library. 59 The Cross at Moone, Co. Kildare. (National Museum of Ireland, Dublin). Courtesy of Visual Arts Library. 64 Hywel Dda. Courtesy of the National Library of Wales. 69 Portrait of Owen Glyndwr, from his Great Seat, engraved from the Archæologia. Courtesy of The National Library of Wales. 70 Rhys Ap Gruffydd, The Lord Rhys. Courtesy of the National Library of Wales. 72 King John Signing the Magna-Carta (Private Collection). 74 Llewelyn Ap Gruffydd. Courtesy of the National Library of Wales. 75 Owain Glyndwr. Courtesy of the National Library of Wales. 77 Treaty of Hengist and Horsa with Vortigern. Courtesy of Hulton Deutsch Collection. 78 The First Meeting of the British King Vortigern with the two Saxon Chiefs Hengist and Horsa. Courtesy of Hulton Deutsch Collection Limited, London. 79 Silver Penny of Ethelbert, Anglo-Saxon King of East Anglia (Private Collection). BAL. 81 Anglo Saxon Monarch, 8th Century. Courtesy of The Fotomas Index Picture Library, Kent. 83 Sigebert, the King-Monk, was Persuaded to Leave his Cell and Lead the Army Against Penda, King of Mercia, by S. Paget. Courtesy of Hulton Deutsch Collection Limited, London. 84 Detail of Illuminated Letter G: King Athelstan (British Library, London). BAL. 88 King Offa of Mercia Overseeing Builder Carrying Bricks (Trinity College, Dublin). BAL. 90 Offa, King of Mercia, Founder of the Monastry at Saint Albans in 793 (Private Collection). BAL. 93 Egbert, King of the West Saxons. Courtesy of The Fotomas Index Picture Index Picture Library, Kent. 94 King Ethelwulf Kneels Before the Pope, Attended by Cardinals, 855, English, with Flemish Illuminations, from Saint Albans's Chronicle (Lambeth Palace Library, London). BAL. 95 Ethelbald. Courtesy of Hulton Deutsch Collection Limited, London. 96 The Viking Sea Raiders, by Albert Goodwin (Christopher Wood Gallery, London). BAL. 98 Edward the Elder. Courtesy of The Fotomas Index Picture Library, Kent. 99 Malmesbury Abbey (Private Collection). 100 Comes Litoris Saxon Per Britaniam: Anglo-Saxon Map, c.950, (Bibliothèque Municipale, Rouen). BAL. 102 King Edgar, Saxon Lady and Page. (Private Collection). 103 St Dunstan Crowning King Edward (Musée Conde, Chantilly). BAL. 104 Anglo Saxon Military Chief, Trumpeter and Warriors, etched by L. A. Arkinson. Courtesy of Hulton Deutsch Collection Limited, London. 105 The Danes Descend Upon the Coast and Possess Northumberland (Wallington Hall, Northumberland). BAL. 106 Vision of the Drowning of King Svein I Haraldsson Forkbeard of Denmark as He Embarked for England (Private Collection). BAL. 108 Cotton Manuscript: King Cnut, (British Library, London). BAL. 109 Harold I: Mary Evans Picture Library. 110 Author Offering His Book to Queen Emma (British Library, London). BAL. 111 Edward the Confessor, by East Anglian School (Richard Philip, London). BAL. 112-13 The Death of Harold, from the Bayeux Tapestry (Musée de la Tapisserie, Bayeux). BAL. 114-115 The Death of King Edward, from the Bayeux Tapestry (Musée de la Tapisserie, Bayeux). BAL. 117 Coronation of William I, from Anciennes et Nouvelles Chroniques d'Angleterre (British Library, London). BAL.118 William the Conqueror receives Allegiance from his Nephew le Roux (British Library, London). BAL. 119 Portrait of William the Conqueror, by English School (Philip Mould, Historical Portraits Ltd, London). BAL. 121 William II, from Historia Anglorum (British Library, London). BAL. 123 Henry I Receiving News of the News of the Drowning of his Soon in the White Ship, by Anonymous (Hartlepool Museum Service, Hartlepool). BAL. 124 Silver Penny of Stephen (Private Collection). 125 King Stephen Enthroned, written during the reign of Edward II, from Chronicle of Peter of Langtoft (British Library, London). BAL. 127 Geoffrey Plantagenet (Private Collection). 128 The Murder of Becket. Courtesy of

Scala. 129 Richard I from The Four Kings of England, from Historia Anglorum (British Library). BAL. 131 Henry II from The Four Kings of England, from Historia Anglorum (British Library). BAL. 132-3 Thomas à Becket departs from King Henry II of England and Louis VII of France (British Library, London). BAL. 135 King Richard I and His Barons, (British Library, London). BAL. 137 King John Hunting (British Library, London). BAL. 138-9 The French King Brought Prisoner to London (Private Collection). 140 Parliament of The Period. (Private Collection). 141 Henry III Being Crowned (British Museum, London). BAL. 142 Edward I with Monks and Bishops, from the Cotton Manuscript (British Library, London). BAL. 143 The Siege of Berwick, by Edward I, 1297, English, with Flemish Illuminations, from Saint Albans's Chronicle, (Lambeth Palace Library, London). BAL. 144 Entry of Queen Isabella (British Library, London). BAL. 145 Marriage of Edward II to Isabella, daughter of Philip IV at Boulogne (British Library, London). BAL. 146 Edward III as founder of the Order of the Garter (British Library, London). BAL. 147 Edward III Granting the Black Prince the Principality of Aquitaine (Private Collection). BAL. 149 Richard II is taken into the Tower of London (British Library, London). BAL. 150 Henry Bolingbroke (Henry IV) enters London, executed for Edward IV (British Library, London). BAL. 151 Coronation of Henry IV, from Froissart's Chronicle (British Library, London). BAL. 152 Marriage of Henry V to Catharine of France, 1461 (British Library, London). BAL. 153 Morning of Agincourt, 25 October 1415, by Sir John Gilbert (Guildhall Art Gallery, Corporation of London). BAL. 154 The Crowning of Henry VI at Westminster, from English Psalter, 1470 (Victoria & Albert Museum, London). BAL. 155 Henry VI of England, by Francois Clouet (Manor House, Stanton Harcourt, Oxfordshire). BAL. 156 Edward IV in Council (British Library, London). BAL. 157 Edward IV, with Elizabeth Woodville, Edward V and Richard, Duke of Gloucester, by Dictes of Philosophers (Lambeth Palace Library, London). BAL. 159 Edward IV of England Landing in Calais, from Memoirs of Philippe of Commines (Musée Thomas Dobree-Musée Archeologique, Nantes). Courtesy of Giraudon and The Bridgeman Art Library. 161 Tower of London Seen from the River Thames from A Book of the Prospects of the Remarkable Places in and about the City of London, by Bob Morden (O'Shea Gallery, London). BAL. 163 Richard III, by Anonymous (Syon House, Middlesex). BAL. 164 Queen Elizabeth I being carried in Procession, by Robert Peake (Private Collection). BAL. 165 Elizabethan London Showing Shipping on the Thames, with Old London Bridge, engraving by Cornelius de Visscher (Guildhall Library, Corporation of London). BAL. 167 King Henry VII, by Anonymous (Royal Society of Arts, London). BAL. 168 Bishop Sherbourne with Henry VIII, by Louise Barnard (Chichester Cathedral, Sussex). BAL. 169 Portrait of Henry VIII, by Hans Holbein (Belvoir Castle, Rutland). BAL. 170 Catherine of Aragon, by M. Sittou (Kunsthistorisches Museum, Vienna). BAL. 171 Portrait of Anne Boleyn, Second Wife of Henry VIII of England, by 16th Century English School (Hever Castle Ltd.). BAL. 172 Edward VI with the Chain of the Order of the Garter, by William Scrots (Richard Philp, London). BAL. 173 The Execution of Lady Jane Grey in the Tower of London in 1553, by Hippolyte Delaroche (National Gallery, London). BAL. 174-5 Queen Mary I, by William Scrots (Private Collection). BAL. 177 Elizabeth I, Armada Portrait, by Anonymous (Private Collection). BAL. 178 James I, Half-Length Portrait, by John the Elder Decritz (Roy Miles Gallery, London). BAL. 180 Anne of Denmark, by Marcus Gheeraerts (Woburn Abbey, Bedfordshire). BAL. 181 Guy Fawkes before King James, by Sir John Gilbert (Harrogate Museums and Art Gallery). BAL. 182 Charles I on Horseback, by Sir Anthony van Dyck (National Gallery, London). BAL. 184 Contemporary Portrait of Oliver Cromwell, by Anonymous (Private Collection). BAL. 185 Battle of Marston Moor, by John Barker (Cheltenham Art Gallery and Museums, Gloucestershire). BAL. 187 Charles II, by Wallerant Vaillant (Phillips, London). BAL. 189 Miniature of James II as the Duke of York, 1661, by Samuel Cooper (Victoria & Albert Museum, London). BAL. 191 Mary II, wife of William III, by William Wissing (Scottish National Portrait Gallery, Edinburgh). BAL. 193 King William III, by William Wissing (Holburne Museum, Bath). BAL. 194 Prince George of Denmark, married Queen Anne of England, 1683, by Michael Dahl (Institute of Directors, London). BAL. 195 Queen Anne, 1703, by Edmund Lilly (Blenheim Palace, Oxfordshire). BAL. 197 Princess Sophia of Bohemia, Mother of George I as a Shepherdess, by Gerrit van Honthorst (Collection of The Earl of Pembroke, Wilton House). BAL. 199 King George I, by Sir Godfrey Kneller (Institute of Directors, London). BAL. 200 King George II, 1759, by Robert Edge Pine (Audley End, Essex). BAL. 201 Caroline II, Queen of George II, by Charles Jervas (Guildhall Art Gallery, Corporation of London). BAL. 203 Bonnie Prince Charlie, by G. Dupre (King Street Galleries, London). BAL. 204 George III & Lord Howe, by Isaac Cruikshank (Guildhall Art Gallery, London). BAL. 205 Portrait of King George III, by Anonymous (Institute of Directors, London). BAL. 206 Westminster Abbey, Coronation of George IV, aqua-tint by F. C. Lewis (Guildhall Library, Corporation of London). BAL. 207 George IV in his Garter Robes, by Sir Thomas Lawrence (National Gallery of Ireland, Dublin). BAL. 209 King William IV, by Sir John Simpson (Crown Estate, Institute of Directors, London). BAL. 211 The Four Generations, Windsor Castle, 1899, Queen Victoria, by Sir William Quiller Orchardson (Russell-Cotes Art Gallery and Museum, Bournemouth). BAL. 212 Victoria: Her Children (Forbes Magazine Collection, London). BAL. 213 Portrait of Disraeli, by Henry Jr. Wegall (Burghley House, Stamford, Lincolnshire). BAL. 214 Nelson's Pillar and the General Post Office, Dublin, by Leo Whelan (Coll. Earl of Mountcharles, Slane Castle, Co. Meath). BAL. 215 The Home Rule Maze – Mr Asquith, 'Excuse me, Sir, but are you trying to get in or out?' Mr Bonar Law, 'Just what I was going to ask you, Sir', by Leonard Raven-Hill (Central St Martins College of Art and Design, London). BAL. 216-17 The New Houses of Parliament, engraved by Thomas Picken, published 1852 by Lloyd Brothers, by Edmund Walker (Guildhall Art Gallery, Corporation of London). BAL. 218 Indian Mutiny and Campaign: Troops Hastening to Umballa, 1859, Engraving from Campaign to India, by G. F. Atkinson (British Library, London). BAL. 219 Imperial Federation Showing the Map of the World, British Empire, by Captian J. C. Colombo (Royal Geographical Society, London). BAL. 221 Edward VII Receiving Maharajahs and Dignataries Prior to his Coronation, by A. E. Harris (Roy Miles Gallery, London). BAL. 223 Thanksgiving Service for George V and Queen Mary, 1935, byFrank Salisbury (Guildhall Art Gallery, London). BAL. 225 Prince of Wales, later King Edward VIII, by Sir William Orpen (Royal & Ancient Golf Club, Saint Andrew's). BAL. 226 HM King Edward VIII, by Walter Richard Sickert (Beaverbrook Art Gallery, Fredericton, Canada). BAL. 227 Edward, Duke of Windsor and The Duchess of Windsor, photographed by Sir Cecil Beaton. Courtesy of The National Portrait Gallery. 229 King George VI, by Oswald Hornby Joseph Birley (Crown Estate/Institute of Directors, London). BAL. 231 Queen Elizabeth II, by Denis Fildes (Institute of Directors, London). BAL. 233 The Crown Jewels. Courtesy of Historic Royal Palaces Photographic Library, Surrey. 241 Map Showing John Speed's 'Kingdome of Scotland', 1662 (National Library of Scotland, Edinburgh). BAL. 243 King Edward the Confessor Seated at a Banquet, from Decrees of Kings of Anglo-Saxon and Norman England (British Library, London). BAL.

Index

Addedomarus, 12
Adelaide, Queen, 209
Adminius, 12
Aedh, 24
Aedh Finnlaith, 23, 56
Aelfgar, 67
Aelfwald, 84
Aelle (of Deira), 85, (of Sussex), 81
Aesc (Oisc), 20, 78
Aescwine (of Essex), 80, (of Wessex), 92
Aethelbert, *see* Ethelbert
Agricola, Gnaeus Julius, 22
Albert, Prince Consort, 196, 210, 211.
Alchred, 87
Aldwulf (of Sussex), 82, (of East Anglia), 84
Alexander I, 32, 33
Alexander II, 36
Alexander III, 37
Alfred, 'the Great', 64, 95, 96, 97
Alric, 79
Anarawd ap Rhodri, 64
Anna, 84
Anne, Queen, 194-5
Anselm, Archbishop, 120, 122-3
Arthur, 21
Athelstan (of East Anglia), 84, (of England), 98, 99
Atrebates, 10, 12, 13

Bannockburn, Battle of, 41, 145
Becket, Archbishop Thomas, 132-3
Bede, 20, 76-7, 81, 85, 86
Beonred, 89
Beorhtwulf, 91
Beornwulf, 91,94
Berhtric, 93
Berthun, 82
Black Death, 147
Black Prince, 146
Boudicca (Boadicea), 13, 14
Boyne, Battle of the, 192
Brian Boru, 56
Brunanburgh, Battle of, 24, 99, 100
Burgred, 91

Cabot, John, 167
Cadell ap Rhodri, 64
Cadwella, 81, 82, 93
Cadwallon, 62, 89
Calais, 147,154, 174-75
Caledonia (Scotland), 18, 22
Calgacus, 22
Camulodunum, 12, 14, 16
Canute, 107, 108
Caratacus, 12, 13
Cartimandua, 13
Cassivellaunus, 12
Ceawlin, 92
Cenfus, 92
Centwine, 93
Cenwalh, 92
Ceolwulf I (of Mercia), 91
Ceolwulf II, 91
Ceolwulf (of Northumbria), 86, (of Wessex), 92
Ceorl (of Mercia), 89
'Ceorl (of Wessex), 92
Cerdic, 92
Charles I, 182-3, 184-5
Charles II, 161, 185, 186-7, 188
Charles, Prince, 'Bonnie Prince Charlie', 202-3
Church, Christianity
Anglo Saxon England, 78, 79, 81, 84, 85, 89, 92; Henry II and Becket, 132-3; Tudors, 168, 169, 172-3, 174, 176-7
Churchill, Winston, 225, 226, 229
Cissa, 8
Coelred, 89
Coenred (of Northumbria), 86, (of Mercia), 89
Coenwulf, 91
Colin (Cuilean), 24
Commius, 10
Commonwealth, 229, 230-1
Constantine I (of Scotland), 24
Constantine II, 24
Constantine III, 25
Constantine the Great, Emperor, 19
Constantius Chlorus I, Emperor, 19
Cormac, 55, 58
Crécy, Battle of, 146
Creoda, 88
Cromwell, Oliver, 184-5
Cromwell, Thomas, 169
Cunedda, 61
Cunobelinus, 10, 12, 13
Cuthred (of Kent), 80
Cuthred (of Wessex), 93
Cynegils, 92

Cynewulf, 93
Cynric, 92

Dafydd ap Llywelyn, 73
Darnley, Lord, 51
David I, 32, 33
David II, 42, 43
Disraeli, Benjamin, 211, 213
Donald I, 23
Donald II, 24
Donald III, 30
Duff, 24
Dumnovellaunus, 12
Duncan I, 26
Duncan II, 30
Dunstan, St, 101, 102
Dunwich, 82-3

Eadbald, 79
Eadbert, I (of Kent), 79
Eadbert II, 80
Eadbert (of Northumbria), 86
Eadric, 79
Ealhmund, 80
Eanfrith, 85
Eanred, 87
Earconbert, 79
Eardwulf (of Kent), 79
Eardwulf (of Northumbria), 87
Ecgric, 84
Edgar (of England), 66, 102
Edgar (of Scotland), 31
Edinburgh, Prince Philip, Duke of, 230
Edmund (of East Anglia), 84
Edmund I, 'the Magnificent', 100
Edmund II, 'Ironside', 107
Edred, 101
Edward I, 38, 39, 40, 41, 63, 74, 142-3
Edward II, 41, 144-5
Edward III, 41, 43, 146-7
Edward IV, 156, 158-9, 160
Edward V, 160-1
Edward VI, 171, 172-3
Edward VII, 220-1
Edward VIII, 224-5, 226
Edward, 'the Confessor', 111, 141
Edward, 'the Elder', 24, 98
Edward II, St, the Martyr, 103
Edward Balliol, 42, 43
Edwin of Deira, 62, 85, 89
Edwy, 'the Fair', 101
Egbert I, (of Northumbria), 87
Egbert II, 79, 87
Egbert (of Wessex), 93-4
Egfrith, 91
Egrith, 86
Elfwald I, 87
Elfwald II, 87
Elizabeth I, 53, 165, 170, 176-7
Elizabeth II, 196-7, 225, 230-1
Ella (Aelle), 87
Empire, British, 201, 215, 218, 228-9
Emrys (Ambrosius Aurelianus), 21
England, Kings of, 6-7, 91, 98, 128-9, 195, 212-13, 216-17, 226, 230-1
Eochaid, 24
Eormenric, 78
Eorpwald, 82
Eppillus, 10
Ethelbald (of Mercia), 89, (of Wessex), 94,95
Ethelbert (of East Anglia), 84
Ethelbert (of Wessex), 95
Ethelbert I (of Kent), 79
Ethelbert II, 79
Ethelburga, 85
Ethelfrith, 85
Ethelheard, 93
Ethelhere, 84
Ethelred (of Mercia), 89
Ethelred I (of Northumbria), 87
Ethelred II, 87
Ethelred (of Wessex), 95
Ethelred II, 'the Unready', 104
Ethelwald Moll, 86
Ethelwalh, 81
Ethelwold, 84
Ethelwulf, 94

Fawkes, Guy, 181
Fergus, 23
Flodden Field, Battle of, 48, 49
France: Scotland and, 35, 49, 50; medieval English kings and, 134-5, 136, 140, 143, 146-7, 150, 152-3, 154, 155, 159; English lose Calais, 174-5; later relations with England, 187, 192, 193, 200-201, 205, 220

Geoffrey of Anjou, 126, 127, 128
George I, 196,198-9
George II, 200-201
George III, 204-5
George IV, the Prince Regent, 204, 206-7
George V, 196, 222-3, 224
George VI, 228-9
Gladstone, W. E., 211
Grey, Lady Jane, 173
Gruffydd ap Cynan, 67, 68

Gruffydd ap Llywelyn, 67
Gunpowder Plot, 181

Hadrian's Wall, 18, 22
'Hallelujah Chorus', 201
Harold I, 'Harefoot', 109
Harold II, Godwinson, 67, 111, 112-13, 116
Harthacanute, 110
Hastings, Battle of, 113
Hengist and Horsa, 20, 78
Henry I, 32, 67, 122-3
Henry II, 34, 56, 57, 68, 69, 70, 130-3
Henry III, 36, 140-1
Henry IV, Bolingbroke, 75, 148, 150-1
Henry V, 152-3
Henry VI, 154-5, 156, 158
Henry VII, Tudor, 61, 162, 164-5, 166-7
Henry VIII, 49, 50, 51, 57, 165, 168-9, 170-1; wives of: Anne of Cleves, 171; Catherine Howard, 171; Catherine Parr, 171; Jane Seymour, 171; Catherine of Aragon, 170; Anne Boleyn, 170;
Hlothere, 79
Home Rule, 214-15
Hun Beonna, 84
Hywel Dda, 65

Iago ap Idwal, 66
Icel, 88
Ida of Bernicia, 85
Idwal Foel, 65
Ieuaf, 66
Indulf, 24
Ine, 93
Ireland, 20, 23, 54-9, 130, 185, 206, 214-15

James I (of England), James VI (of Scotland), 52, 53, 179-81
James II (of England), 188-9, 192
James I (of Scotland), 45
James II (of Scotland), 46
James III (of Scotland), 47
James IV (of Scotland), 48
James V (of Scotland), 50
James VI, see James I (of England)
John, 36, 71, 73, 136-7
John Balliol, 38, 39
Julius Caesar, 12, 16

Kenneth I, Mac Alpin, 23, 24
Kenneth II, 25
Kenneth III, 25

Llywelyn ap Gruffydd, 74
Llywelyn ap Iorwerth, 'the Great', 71-3
Llywelyn ap Seisyll, 66
Ludeca, 91
Lulach, 27

Macbeth, 26
MacMurrough, Dermot, 56
Madog ap Maredudd, 68
Mael Sechnaill III, 56
Maelgwyn Hir, 62
Magna Carta, 73, 137
Malcolm I, 24
Malcolm II, 25
Malcolm III, 34
Malcolm IV, 34
Maredudd ap Owain ap Hywel Dda, 66
Margaret of Anjou, 156, 158
Marlborough, Duke of, 191, 194, 195
Mary I, 170, 173, 174-5
Mary II, 190-1, 192
Mary, Queen of Scots, 51, 53
Matilda, 123, 124-5, 126-7
Mayflower, 180
Merfyn Frych, 63

Niall of the Nine Hostages, 55
Normans, 56-7, 111, 114-15, 116
Nothelm, 82
Nunna, 82

O'Connell, Daniel, 207
O'Connor, Cathal, 57
O'Connor, Rory, 56-7
Octa, 78
Offa (of Essex), 81
Offa (of Mercia), 84, 90
Olaf Guthfrithson, 100
Olaf the White, 23, 56
Osbald, 87
Osberht, 87
Osmund, 82
Osred I, 86
Osred II, 87
Osric, 86
Oswald, 85-6, 89
Oswini, 79
Oswulf, 86
Oswy, 84, 86, 89
Owain Glyndwr, 61, 75
Owain Gwynedd, 69

Parliament, 216-17; medieval, 129, 141, 143, 147, 151, 153, 158-9; Henry VIII, 169; Stuarts, Civil Wars, 178, 181, 182-5; revolution of 1689, 193; Catholic Emancipation, 207;

Reform Act (1832), 208-9
Penda, 62, 84, 89
Plantagenets, Angevins, 127, 128-9
Prasutagus, 13
Prince of Wales, 61, 143
princes in the Tower, 160-1
Pybba, 89

Raedwald, 82
Regni, 10
Rhodri Mawr, 63
Rhys ap Gruffydd, 'The Lord Rhys', 70
Rhys ap Tewdwr, 68
Richard I, 35, 134-5
Richard II, 148-9
Richard III (Richard of Gloucester), 156, 160, 161, 162-3
Riesig, 87
Robert I, le Brus (Bruce), 38, 39, 40-1, 145
Robert II (Stewart), 43, 44
Robert III, 44
Robert Curthose, Duke of
Normandy, 120, 122
Robert of Gloucester, 126, 127
Roses, Wars of the, 156
Rothesay, Duke of, 44
Rye House Plot, 187

Saebert, 80
Saelred, 81
Scone, 27, 142-3, 234
Scots, Scotland, 18, 20, 22-53, 99, 14-3, 145, 150, 185; Act of Union, 195; Jacobites, 202-3; George IV and, 206; Home Rule, 215
Seaxburgh, 92
Sebbe, 80
Shakespeare, William, 163, 176
Siegeberht (of East Anglia), 82-3
Siegeberht (of Wessex), 93
Siegeberht the Good (of Essex), 80
Siegeberht the Little, 80
Sigered, 81
Sigeric, 81
Sighere, 80
Simon de Montfort, 140-1
Sledda, 80
Stephen, 33, 124-5, 126-7
Strathclyde, 22, 24, 25, 77, 100
Strongbow (Richard de Clare, Earl of Pembroke), 56-7
Swafred, 81
Sweyn, 'Forkbeard', 104, 106
Swithhelm, 80
Swithred, 81

Tara, 55, 58
Tasciovanus (of the Catuvellauni), 12
Tasciovanus (of the Trinovantes), 12
Tincommius, 10
Togodomnus, 12
Tower of London, 161
Turpin, Dick, 201
Tytila, 82

Ui Neill, 55, 56
Ulster, 54, 55
United Kingdom, 195

Verica, 10
Victoria, Queen, 196, 210-211, 212-13, 218
Vikings, Norsemen, Danes, 55-6, 77, 93, 94, 95, 96, 97, 104, 105, 107
Vortigern, 20
Vortimer, 20

Wales, 12, 18, 60-75, 90, 99, 142, 143, 215
Wallace, William, 39
Walpole, Robert, 198-9, 200
Warwick, Richard Neville, Earl of, 'the Kingmaker', 154, 156, 158, 159
Wiglaf, 91
Wihtred, 79
William I, 'the Conqueror',
Duke William of Normandy, 28, 68, 111, 113, 116, 118-19
William II, Rufus, 31, 120-1
William III, of Orange, 189, 190-1, 192-3
William IV, 208-9
William I, the Lion (of Scotland), 35
Woodville, Elizabeth, 159, 160, 162, 165
Wren, Christopher, 190
Wuffa, 82
Wulfhere, 89

York, Richard, 3rd Duke of, 154, 156
York, Treaty of, 36